Original front cover from 1917

In Naval Airship Against England [1]

By: Over Lieutenant at Sea
Baron [2] Treusch of Buttlar-Brandenfels

Translation and Commentary: Madison Hasson

11 to 20 Thousand

1917

Eckhart-Verlag, A.-G., Berlin SW 68

All rights, especially that of translation reserved. [3]

American Copyright by Eckhart-Verlag, A.-G, Berlin

. * . * . * .

1 - Yes, That's the title. Germans like long descriptive titles. You should see their title for the movie Airplane…"The unbelievable trip in a crazy airplane!"…. It doesn't work very well in English, but in German it's an awesome title. If I were to rename this for an English speaking audience it would be "My Zeppelin VS England".

2- A more literal translation of the German title usually translated as Baron is Free Lord. As in a Lord who legally owes no fealty or taxes to anyone else. A Lord who is free.

3 - Note: This book is so old it is now public domain. This is the first English translation. I maintain the copyrights for this translation, but not to the original book.

Graf Zeppelin und der Führer der
Marineluftschiffe, Korvettenkapitän Strasser.

Count Zeppelin and the Leader of Naval Airships, Corvette Captain Strasser.

An airpatrol attacking an airship. (English Depiction)

Meine Besatzung.

My Crew

Graf Zeppelin an Bord die Karte studierend.

County Zeppelin on board studying the map.

Ein Geschwader wird passiert.

Passing a squadron.

Schwenkung eines Geschwaders vom Luftschiff aus gesehen.

Squadron turning as seen from an airship.

Se. Exzellenz der Flottenchef, Admiral Scheer, vor der Führergondel des „L. 111".

Senior Excellence of the Fleet command, Admiral Scheer, in front of the lead gondola of "L111"

Im Kampf mit Luftschiffen,
die die deutschen U-Boote begleiten.
(engl. Darstellung).

In battle with airships accompanying German U-boats. (English depiction)

Ein englisches Wasserflugzeug steigt auf, um einem Luftschiff den Rückzug abzuschneiden.
(engl. Darstellung).

An English seaplane taking off to cut off the return of the airship.
(English depiction)

Forward

The request of this publisher, to write a small book over the activities of our Naval Airships in the current great war, I am happy to follow. Even this, our youngest weapon has so many excited followers, young and old in this people [1], that it seems worth the effort, as far as military interests will allow, to report something ongoing about them, especially since that which has recently been published does not impressive those who have experience with this weapon system.

How often have I been attacked with questions, probably and improbably, from people who want to know something about the Naval Airships. How often have I heard, "Yes, the activities of our torpedo boats and submarines we know every last bit, but we never hear anything about the airships." Now that should be the purpose of this booklet, to at least give the reader an approximate picture of the activities of the Naval Airship Program, how they actually are, the weapon, who next to her sister, the submarine, has a special calling in this ongoing World War, which is to beat our greatest and most dangerous enemy's hide raw [2]; and this should possibly help to bring about our final victory over our enemies.

So with joy, but not completely without anxiety [3], I started writing this booklet, because - it must be written; and I am not a poet by profession. My last great work, that I put together, was withheld from many distant circles and was a pretty long time ago. It is an essay from highschool [4], it cost me a lot of sweat…in spite of that, I will try the quill again!

The subject back then - in school - was not anywhere near or dear to my heart; but over today's subject, I know a bunch of things to talk about. [5]

What I tell here, is mostly things I've seen with my own eyes, that I've experienced since the start of the war with my own brave crew on several different airships. But in some spots I have weaved in experiences of my ship mates [6] that seemed noteworthy to me.

If the reader at the end of this booklet lays it down, knowing they have experienced and learned something new about airships, and if they get a new interest, even excitement for this new weapon, then my purpose has been completely fulfilled.

-The Author [7]

.*.*.*.

1- The Germans

2 - This was an unfamiliar idiom for me. The literal German phrase was "to give them roughness on leather". This was the closest approximation I could think of that would work for an American audience.

3- The German word here more directly translates to shy, but with the way it was used here, I almost went with nervousness, but settled on anxiety. A hundred years ago, I believe shyness would have been just as accurate a translation, but we've stopped saying shy and started using social anxiety instead.

4- This would have been somewhere around 1904-1906 the period

5-The nice thing about being an adult is that if you choose to write anything it is generally something you want to write about. This by no means excuses students from writing things because they don't want to. These school essays help you learn to write so that when you choose to write, you aren't starting from scratch..

6- Ship mates is the US Navy's term for brothers in arms. The Germans use the word comrade. Due to the communist revolution in Russia and the cold war, the word comrade had too much cultural baggage to use for an American audience. I believe the term "ship mates" stays truer to the German usage of comrade during WW1.

7- In case it isn't clear, the author is Over Lieutenant at Sea Free Lord Treusch von Buttlar-Brandenfels.

Translater's Forward

This is my second book to translate. At first I thought it would be a fascinating account of zeppelin stories. It was, but I believe this book has more value than as just another account of zeppelin use during WW1. This book was a fascinating window into the German cultural consciousness during WW1. It is clear that they believed they would win. It also seemed to me that this book was also meant to make American audiences wary of joining the war. This book was published in German in the United States, in 1917! I knew there used to be over 3000 German language periodicals in the United States before the World Wars, but it never quite hit me until I read this book. That the Germans would make use of these publishers to woo and warn the American population makes way too much sense. I have made some format changes. In the original book the table of contents was between the end of the last chapter and the advertisements. I have moved it to the front of the book which is more standard for American audiences. I have also added a couple of reference sections to the book and have included them into the table of contents. In the original book, all of the pictures were interspersed throughout the book, with several at the beginning. I put them all at the beginning of the book. I also translated the advertisements at the end of the book. Not that I intend to translate these books…but I might in the future, but because I think it provides cultural insights into the Kaiser's Germany, that I don't think you could get any other way.

    When translating this, I was hoping to get some insight into airship operations, since I intend to build and operate airships during my lifetime. The books I translate are translated with the hope that book sales can be used to fund construction of airships. If book sales are low, at least I've added to the amount of information available to the English speaking world from World War 1. I hope you enjoy this book. Thank you for buying it.

    -The Translator
    -Madison Hasson PE

	Page
Pictures………..	…..2
Forward……………	…..11
Translater's Forward………..	…..13
1. Requisition and Transfer……	…..15
2. Christmas on the North Sea…	…..20
3. The Trip in the Woods……	…..27
4. Attacking London……	…..35
5. Battling English Submarines…	…..48
6. Traveling in a Storm!……	…..53
7. Encounter with English Sea Forces……	…..63
8. Scouting with the Fleet…	…..69
L130……[1]	…..84
Advertisements …….	…101
Imperial German Rank Structure……	…106

[1]- This section did not have its own chapter in the original German, but it was included in the table of contents as its own section. I will still treat the footnotes/commentary for that "not" chapter as the footnotes/commentary for the other chapters.

Chapter 1

Requisition and Transfer

It was late morning in autumn 1914, I was with a ship mate at the Hamburg airship hangar, pacing back and forth, graciously swearing at the bad weather that had destroyed any chance we had at "traveling", when the telephone operator rushed in with a message for me: "The LoA [1]. Is on the phone and wishes to speak with Mr. Over Lieutenant."

Naturally I had a guilty conscience, that I deserved something and was just about to get my just desserts [2]. I hurried to the phone, not too happy, resumed my military bearing, like anyone involuntarily does whenever they hear their commanding officer on the phone, and report on the spot. How happy I was to find out I wasn't getting any just desserts. Quite the contrary! I was told that I was selected to command the L66, and at this and that time I should go down to the Zeppelin factory in Friedrichshafen and start my command.

My training as a student and watch officer was over! As a commander I am receiving my own airship!

In the few days I had in Hamburg, before my vacation in Friedrichshafen to take my ship, all my work was given to my successor so that everything would be taken care of till I could take my command. He left me with the best wishes for my future ship. We celebrated at the small cantina on the airship field, at the insistence of the old and young "W.O."'s (watch officers), and that evening I traveled happily and peacefully South.
I don't rightly know if my vacation days went by too fast or too slow. Vacations are nice - but I longed to be in Friedrichshafen where my ship was already being built.

The Watch Officer and the crew were already down there. I shared my arrival with them and first got to know Ferryman at Sea von Schiller, who was ordered to serve as my Watch Officer, who even today still goes through thick and thin with me. [3]

The next few days we went to the shipyards, where we saw for the first time, with gleaming eyes, our ship.[4] They had already made a lot of progress on the ship. The skeleton already stood complete in the hangar. If everything went well, the ship would be finished enough to take in eight days. The days to completion went quick. The day of the test flight had arrived. The ship left the hangar, and with his Excellency, the Count von Zeppelin, on board, the take off led into a successful first trip. Everything went smooth, not even the smallest bit of drama [5], so I could report that same day, that the requisition went through without any difficulties and that the inaugural flight to the port of destination would happen tomorrow.

At 4 am the next morning, L66 sat in its hangar ready for flight. The construction workers walked the ship to the field. Then we bid farewell to all the friends we had made during our short stay by Lake Constance.

The old Count did not get up and join us, but sent a speech for one of his employees to read. It was so moving, deeply heartfelt, and energizing, that we all cheered ,"Hurrah!" for the new ship. As a token of our gratitude, the crew answered back with a mighty, "Hurrah!"

The engines received the order to start. The ship raised majestically and started its journey to the front, accompanied by the most fervent wishes for its future.

The ship quickly disappeared in the dark as it headed north. This flight was the real test flight. So instead of heading straight north, our flight meandered across all of Germany.

In the morning twilight, we toured over Swabia. The first beams of sunlight hit Geislingen in its valley, then we went on to Stuttgart, Karlsruhe, Heidelberg, Darmstadt, Frankfurt, Mainz, and finally to the Rhine valley onward to Cologne. We were supposed to arrive there at dark.

Our path went onward to Bremen. We were already really close to the hangar that would eventually receive us, but we still had a lot more places to visit. Tomorrow we were supposed to visit Hamburg, Schwerin, and central Germany. About 10 AM Berlin came into view. You couldn't see any houses or anything like unto it…just a thick layer of smoke. [6]. Many of the crew thought about them below us, "Thousands of people live in this smokey, dusty place. Those poor devils!". When we got over the actual city, we started to be able to see the houses. The people stood and waved at us with their hats and handkerchiefs.

We were supposed to reach our home port in the evening. From Berlin we traveled over Stettin, where the Baltic was beautifully clear, then towards Frankfurt on the Oder, back over to Rathenov heading towards Hamburg, onto the mouth of the Elbe and finally to our homeport. [7] We got there in the late afternoon. The ground crew was already on deck, ready to catch the ship. The ship drifted slowly down to the landing. As soon as the ground crew got a hold of the mooring lines, the band greeted the ship with a lively, joyful march. It was how they wished us luck.

L66 arrived successfully. It had passed its test flight with flying colors and was brought into its protective hangar. This ship brought new capabilities to our fleet's front line. In a short time, the ship and crew would have the chance to prove themselves against the sea and the enemy.

1- Leader of Airships, in German it is Führer der Luftschiffe (FdL), let it be noted, this was in 1914, long before fuehrer became associated with a national socialistic genocidal psychopath, so this has no such connotations that Fuehrer conjures up in the modern English speaking mind, when this was first published. German has two words for Leader. Both are based on verbs, both translate roughly to "leading", but there is a nuanced difference. Fuhren is the verb to describe what a shepherd does. A fuehrer is a leader that has been there in the trenches before, loves the people, knows what the people need, understands how their work is done, and leads by example. The other word leiten, pronounced like the English word lighten, is the same thing but without the heartfelt love/respect between the person leading and those following, more like a callous business manager, who has no clue what those under him have to do to get things done, and doesn't realize or care how his actions make their lives harder. There's a reason certain Austrian Corporals chose to use the word Fuhrer as their title. They wanted the German people to see them as a loving shepherd.

2- The idiom in German is "catch one on the cold"... It doesn't translate well directly, but it is a euphemism for when you bop your dog on the nose for bad behavior.

3- I used an American idiom. In German he just says "shared much joy and pain with me."

4- I imagine this unfolded much like a scene from spongebob, where spongebob and patrick are joyously staring with sparkling eyes at a brand new vehicle just for them. Imagining a couple WW1 Germans in full uniform doing the exact same thing just makes me laugh. But if you were just given command of a zeppelin, you'd probably do the exact same thing. I know I would.

5- The word used in German literally translates to display.

6- The thick smoke was due to the fact it was winter, and in the early 20th century almost everyone heated their homes by burning wood or coal, especially the Germans. Coal was the most advanced wonder fuel at the time. It had a much higher power density than wood or charcoal, and engines that used gasoline or diesel had just been invented. Germany had (and still has) huge natural reserves of coal. It was how they powered the industrialization of the German Empire. It was such an important resource that the black on the Imperial German flag stood for the coal that Germany's power was built on.

7- Approximate path of the maiden voyage as described in this chapter:

Chapter 2

Christmas on the North Sea

Christmas Eve! The whole crew of L66 were gathered around the Christmas tree in the warrant officers club having a Christmas party. Everyone was enjoying the chatting, smoking, & singing. Then someone brought in an order from the Airship Admiral, "Be over Helgoland when day breaks. Then head North and observe."

The grog is finished off, now, without conversation. The only thoughts now are of getting a couple hours of sleep before take off.

About 5 AM, the ship was cleared for take off. During the night, snow had covered the whole landscape. With a light wind out of the North and no clouds, it got very cold. The temperature was -5 C [1]. The most beautiful weather for airship flights.

The wind and weather reports were better than good. We could get this thing started.

In the hangar that housed L66, everyone was diligently working. The engines were started, so they would be warm and work better in the cold. Many less than friendly words drifted through the air as they tried to start the engines running in the cold. Once the engines started, everything was good. All sorts of things were brought onto the ship: gasoline tanks were filled, ammunition for our artillery [2], bombs in their various sizes and weights, provisions for our bellies, and many other things were brought onboard. The ship was supposed to be ready for lift-off at 6 am.

Shortly before the appointed time, the ground crew, "the parkour boys" [3], our joking nickname for them, arrived to take the ship out of the hangar. The Watch Officer weighs off the ship, meaning he lets out water ballast until the ship starts swimming.

"What's that supposed to mean?", the reader may ask. When the ship is "weighed off" by draining off water ballast until the ship is so balanced, or as the seaman would say, "trimmed", that it floats. It doesn't try to go up, nor does it try to rest on the gondola. When a ship is weighed properly, it only takes one man to push the ship into the air. It always makes a huge impression on our visitors, when they see one man lift the ship.

At 6 AM, on the dot, the Watch Officer messages, "Ship cleared for departure."

Upon the command, "Airship March!", the ship glides out of the bright lighting in the hangar out to the field. There, it is positioned so it can climb without hitting anything. Then the mooring lines are cut. * (* The engines idle. The ship is cleared for take off.)

"Both rear engines, full ahead!"

The engineer's telegraph transmits the orders to the rear gondola. The propellers start, first slowly, then gradually turning faster, until all you could see was a circle glistening in the morning sunlight. The ship starts to move forward.

"Up!"

On this order, the people who had until now been holding tightly onto the gondola, throw the ship with all their might into the sky. The bow rises, and slowly the ship climbs into its element.

"All engines full forward!"

Now all the propellers are clutched in, and the ship picks up speed, until it rushes through the air at about (redacted) kilometers per hour.

The order sounds, "Set course for Helgoland!"

The helmsman sets the course accounting for the prevailing winds on our weather map, and orders the under officer on the side rudder, "NtE ½ E" (meaning North to East half East). The ship slowly turns towards its new heading until they report back, "NtE ½ E, course set."

A quarter hour after take off we reach the coast. In this crystal clear night, you could clearly see for miles the dark strip where land and water meet. Just outside the mouth of the river, we passed over countless fishing boats and outpost ships.

When it was light, we were at Helgoland. The sun rose over the Eastern horizon as its light streamed over the North Sea as it lay before us.

As morning daylight increased, so did our visibility. The port of Helgoland was bustling. Some torpedo boats and submarines were leaving, some were returning from their night patrols. Countless sea planes floated on the water, ready for take off. Some were already in the air, circling around our ship, following our course and disappearing over the North Sea

"Four clicks ahead to port is an airship", reported the look out on the platform through the speaking tube.

You couldn't see anything with the naked eye. With binoculars you could see a gray stripe, L50, that took off this morning at the same time as L66 and was traveling and the reconnoitering the north-western sector.

Helgoland disappeared from view. Far and wide, there was nothing to be seen. Suddenly we noticed L50 had altered course and was climbing.

Three airplanes headed directly at them had come into view. We changed course to observe what type of people they were! The utmost caution was ordered.

With the great speeds airships and airplanes have, it didn't take long for both parties to close the gap and be able to identify their type. They were enemy airplanes headed our way. The under officers on the machine guns, in the rear gondola and on the platform, were cleared to give these English gentlemen a proper greeting.

The commander issued the order, "fire at will." The command went out from the forward gondola to all the artillerists. The airplanes had come within gun range.

Just as the report of the first shots from L50's machine guns sounded, the three airplanes scattered. But they didn't retreat. They had simply split up.

L50 couldn't keep up with the planes, they still had reconnaissance to do. So they reported the airplanes. "Passed three enemy planes. Their course was South East."

We on L66 also picked up their radio message. It lifted our spirits. There was activity and excitement to be had over the North Sea today. The enemy didn't bring their ships into German waters, so they were probably trying to figure out something with their planes. Now we could explain why L50 suddenly changed course and altitude. The lookout was told to keep their eyes peeled.

After a little while, we saw smoke in the North. Was it an enemy ship? It would be too wonderful if we found them, but no one believes it's enemy ships. How many times had smoke just been old steamers and fishing boats hard at work?

But this time our hopes weren't in vain. This Christmas gave us a couple of real English ships! When they came into view, they were a small cruiser and torpedo boat, each with their own steamer. They were slowly headed west.

We immediately sent out a report. As L66 approached them they increased their speed. They had no chance of running away. They didn't know how fast we could go.

"I think those bloke's are shooting at us," the helmsman stated as he watched the ships through binoculars. Sure enough, we could see the white gunpowder smoke wafting in the air. Everyone looked out the window to watch the display. "If they keep shooting like that, they'll never shoot us down!"

So we chased the ships, headed west. They had reached their maximum speed. We hung over them, stuck to them like a burr. They couldn't shake us off. A few clouds came in from the Northwest. They came at just the right time. One of the steamers was lagging behind. We didn't know what was wrong with it. It became our next target.

L66 climbed higher into the clouds. The ships below could hear the drum of our engines, but had no idea where we were. In the meantime, the bombs were armed and cleared. When we dropped below the clouds we were almost directly over the steamer. They became very uncomfortable and started to steer around like a drunkard. It was the most amazing zig zag pattern, it was his only defense against our bombs. I could easily think of a hundred situations I'd rather be in than the one that steamer was in.

The Watchofficer stood at the targeting scope, ready to drop the bombs. Slowly, L66 approached its target. From up here we could see how the men were standing on deck and shooting at us. We never thought they could actually do anything to us.

There! Bombs away! The first bomb falls. It landed short, about 30 meters [4] from the steamer.

Now our work really starts. Our own planes have arrived and are attacking with us. We observed a hit on the middle of the ship.

The cloud cover gets thicker and hides the steamer from our sight. Now, what to do?.... We still have our artillery. [2]

We started a steep dive to break through the clouds to see how thick the cloud cover was. The clouds were at 500 meters [5] altitude, even at 300 meters [6] altitude. At 250 meters [7], the cloud cover thinned, at 200 meters [8], were out of the clouds and right next to the steam boat. Immediately our machine guns open fire. Suddenly, all those people standing on deck that were shooting at us, desperately disappear from the deck.

The enemy's small cruiser, noticing what was happening, charged our way at top speed and opened fire. Unfortunately, we had run out of ammunition. There was nothing more for L66 to do. Before the enemy fire got close enough to worry us, we were already high in the clouds.

On the way back home, L50 found some planes floating on the water. Maybe they were the planes we saw a few hours earlier. They had been abandoned by their crews and we're a welcomed prize for the torpedo boat that brought them in.

L66 landed in the late afternoon. The ship had become heavy. Per our calculations we had to drop ballast. But first they had to clear a spot for us. Our ship had actually been hit a lot. It couldn't have taken down the ship. All it did was keep the sailwright busy. He had to fix the holes. He didn't mind because they were "ideal" holes. After a couple hours, our bird was whole.

Now we heard from our shipmates the whole story of L50's adventure. L50's radio message was received in time for the planes to find our defensive batteries armed and ready. Their Christmas visit to the North Sea coast had gone miserably. One of the planes made it over land. Our batteries unloaded on him. He ended up dropping his bombs in a meadow and heading back. Later, he was rescued from his plane by a fishing boat.

I don't know if those two holes in the ground in a meadow are still visible today[9]. Was the entire endeavor worth losing four planes? [10][11]

.*.*.*.

1 - 23 F

2- Germans considered machine guns a type of artillery. Airships had multiple machine guns on board, no artillery cannons.

3- The word used in the original German was a French term for acrobats that perform on the ground. I translated it to an American term for people with similar skills that is in common usage in the very early 21st century.

4- Roughly 100 ft

5- About 1640 ft.

6- Roughly 1000 ft

7- About 820 ft

8- About 656 ft

9- 1917, Those holes are not visible in 2022. I doubt they survived WW2 with how much the allies bombed Germany.

10- This book was published in 1917, while the war was still ongoing. This question of, "is it really worth it?" was one the Germans asked of the British quite frequently during WW1. I suspect this book, published in German in the United States, was probably a propaganda effort by the Germans to keep America neutral.

11- Another account of this event can be found in Knights of the Air in the chapter titled "Attacking the Aether". I translated that book before this book. It is available at Amazon.com.

Chapter 3

A Trip In The Woods

The telephone rang.
The LoA [1] is on the horn and gives the following orders:
"An seaplane hasn't returned from his reconnaissance flight northwest of Helgoland. Take off immediately and search."

This meant to leave as quickly as possible to find the plane that went down, probably from engine trouble, that was bobbing around in the North Sea and help him. Alarms went off for the crew and ground crew. The hangar doors were opened, and in a quarter hour, L66 was heading towards the North West.

The weather was not very inviting. There was a Southerly wind. Light rain showers fell from time to time. The barometer was falling. We couldn't predict what the weather would do. It could stay the same. It could get much worse.

Before take off, I asked my former instructor, Dr Eckener, who had been traveling with Count Zeppelin for many years, and had a lot of experience predicting weather. He was somewhat of a tree frog, and I sought his counsel. [2]

"What do you think about the weather, Mr. Doctor?"
"Yes," he answered in his Schleswig-Holstein accent, "I have no doubts, there is a small tornado somewhere over the North Sea. You can travel just fine, but watch out for sudden southerly winds."

With the wind, the ship made an exemplary flight over the sea and reached and passed Helgoland rather quickly. Now we just needed to search for the missing plane in our assigned area.

Many torpedo boats were in the area. Till 4 pm, we found nothing. The weather became more and more unfriendly. Thick clouds gathered to the south and southeast. The wind picked up, and the sea continually had white cap waves. Sometimes torpedo boats are swallowed by such seas.

The decision to head home was made easy, because it had started to get dark. Just as the order was given to return, a boat reported they had found the plane, the two occupants and the plane were safely aboard. We could travel home without worrying about them. Too bad we didn't find them. It would have been nice for the "overblown competition", that's what the plane pilots jokingly called us, to have something to hold over them.

L66 turned South East and headed back to Helgoland. In this wind we could make it home by 7 pm.

"Mr. Over Lieutenant, we're getting more wind," reported the helmsman. "We're barely moving forward."

"It doesn't really matter," replied the Watch Officer, "if we land one or two hours late, our dinner is still going to be cold. If we were on time, tonight's dinner would still be served cold."

Actually, the wind picked up and started to really push back on us. We climbed in altitude to see if there was less wind up high. Sometimes it's very blustery down low, but perfectly still up high. Unfortunately that was not the case. There was more wind up high. So we went back to 200 meters [3] altitude.

The ship started to bounce. The weather had become gusty. Pretty soon, it started to rain. The watch on the platform reports to us down below, "it's raining hard.". The storm Dr Eckener had predicted was here and it was really crappy weather.

Then the platform watch reported, "It's starting to snow!". Soon all the windows, and even the tension wires in the gondola had a thick crust of ice. The gusts became more frequent and stronger. The ship was quite literally thrown around. The signal mate on the ruddervator controls had to pay extra close attention to make sure the ship didn't go too low, or we'd all end up in the water. He had to crank hard on the wheel to keep the ship at altitude. But he was immediately rewarded for his efforts. He was the only person comfortably warm. Everyone else froze like a poor, homeless beggar.

My main concern was the engines. If one broke down, things would be pretty precarious. But engines rarely ever broke down just one at a time. Usually a second engine would break down to keep the first one from feeling lonely. I had the lead machinist come to the forward gondola, instructed him to make sure we have no engine breakdowns and asked how the engines were running. His response was calm and very reassuring, "All engines are operational, Mr. Over Lieutenant. They are all churning away like coffee grinders."

Actually, the engines did a fine job. None of them broke down. Even though the gusty weather was taxing for them.

Ahead, we could see lights. That was probably Helgoland, and after a long while the island was far behind us. Our visibility deteriorated. The rain, which was intermediate at first, ceased to stop. To make things worse, snow had accumulated on top of the ship, the rain, snow, and hail had frozen to the ship, adding a lot of extra weight. The rain dripped into the gondolas. There was so much water, our maps and other papers were floating. Eventually the lights on our compass and the ruddervater's instruments went out. To put it simply: the weather was getting much worse.

The lights of Helgoland slowly wandered away behind us, but just wouldn't disappear. The wind was blowing just as fast as we could move forward. Our forward movement was at a snail's pace. We must have reached the coast some time around 8:30 pm. Since it was so dark from the rain and snow flurries, we couldn't see the coast. Every five minutes, we dropped incandescent bombs to see if we were over water or land. It was water, water, and more water. But the engines were still running without a problem. We were moving slowly, but we would still be getting home safely.

About 9:15 pm, one of the incandescent bombs landed in a mud flat. We knew we were getting closer to the coast. Now we had to figure out where on the coast we were. It was easy to get turned around in the wind and drift off our intended course. A well known guiding light, that is kept burning near the mouth of the Weser, was spotted. We knew how to get home from there.

We arrived about 10 pm. They had everything ready for us. The ground crew was ready. The hangar was open and well illuminated. Those below had started to worry, since it was over two hours since our last radio transmission. The ground crew ran to meet us.

We saw the wind speed and direction and looped around. Airships need to land flying into the wind. Naturally, the wind blew the ship away, causing the landing field and hangar to disappear. It didn't take long for us to get back to our landing zone. We held our altitude at 175 meters [4], when suddenly a gust grabbed the ship and threw it down. The ruddervator man immediately ordered, "Rudder Up," but it didn't help any more. There was a sudden jolt followed by a loud crack a few seconds later. We had hit something.

Everything is pitch black. The forward engine had been shut off, so it was very quiet in the front of the gondola. I noticed someone's hand dart across my face and catch a fur tree branch.

We were in the forest.

The sudden jolt had changed the forward gondola quite a bit. The steering wheel, maps, compasses, instruments, thermoses, and everything else had fallen on the floor.

I was a bit speechless. The helmsman, who was standing next to me and also saw the entire scene unfold said, "Wow, Mr. Over Lieutenant, that got us good!"

We had all kinds of problems. Many we couldn't do anything about. When you fly your airship into trees at full speed, you're lucky if you have anything left. The situation was so bad, it was funny. The helmsman, an experienced sailor from Hamburg, couldn't get over it. He got out the log book, wrote down the exact time of our landing, if you could call it that, and the other things like air temperature, gas temperature, and whatever else was required to be written down whenever we landed. The ship was pretty well anchored in front. As the lift gas leaked out, the tree's sank deeper and deeper into the hull, preventing the wind from catching the ship and taking it back out to sea.

At this point, I really wanted to know our location, so I could make a telephone call to the airship field and ask for help. A light appeared below us, so I hollered out, "Hello! Where are we?"

"In the forest," came the reply.

"I already realized that!" I responded.
"Where did you come from?" I probed, hoping the answer would tell me which hamlet we were near and how far we were from the airfield.

"Out of the rear gondola! Mr. Over Lieutenant!" answered the voice out of the darkness below us.

Everyone in the forward gondola couldn't suppress their giggles. It was a machinist mate from the rear gondola. When the rear of the ship got caught, he had jumped out and ran ahead to try and help the front of the ship.

The Watch Officer and several Under Officers climbed out, lowering themselves with ropes or climbing down trees to figure out where our ship was.

In the meantime, people from the local village had started to gather. They had heard the ship and we're wondering why the engines stopped so suddenly. Some had heard the crash of the ship hitting the trees. With their help, we figured out where we were.

Turns out we were really lucky. Only the nose of the ship, from the forward point to the forward gondola, was stuck in trees. The rest of the ship rested on an empty meadow. Most of the ship was undamaged, and what was better, was that the damage we did have was only on the bow of the ship.

Just as quickly as the storm came, it stopped. Right after we got stuck in the trees, the rain stopped, the wind died down, the clouds disappeared, and we could see the starry night.

We couldn't stay there forever. We had so many people show up on scene: cops, professors, women, girls, and especially boys. All of them were offering their help. So I decided that with their support, we would try and move the bird to a more comfortable situation.

Every expendable crewman was sent down to help. That was enough weight that the ship started to float again and with a few loud cracks freed itself from the trees and branches. Then we moved the ship back so that it was away from the trees and rested safely on the meadow.

We were supposed to be 6 kilometers [5] from the landing field, but the locals couldn't agree on which way would be most direct. Over this way was a bog, over that way were houses, and all other sorts of things that were not helpful to moving a sick airship back home.

In spite of these obstacles, we marched onward. After all, we had to at least try. Men, women, girls, and boys, anything with legs grabbed onto the mooring lines and moved the airship. [6] One man volunteered to run to the nearest phone and call the airship station and inform them of what was going on.

We had been walking for about twenty minutes, when hundreds of lights appeared in the forest. It was the ground crew coming to meet us armed with their flashlights. A few hundred men were now helping and the trip went faster than I thought possible considering the darkness and uneven terrain.

There were still many obstacles to overcome. The railway embankments, telegraph lines, and phone lines were especially difficult. But we handled it with grace, and once we got over the fence of the airship fields, we knew the ship was safe.

L66 was taken into the hangar. Everything was already prepared to carry out repairs. The repairs started immediately.

In just five days of very strenuous work, the ship was ready to fly. The test flight was satisfactory.

During the next few days, everyone in the crew made a pilgrimage to where the ship got stuck in the trees. It was fascinating to look at. Tree trunks about 25 centimeters [7] in diameter weren't strong enough and had been snapped clean off.

Even today [8] you can still see where the airship got caught in the trees. Now whenever we fly over that spot, someone always jokes, "Don't forget to rudder up!"

1- Leader of Airships

2- In Germany, tree frogs disappear before really bad weather, so they can be used to predict the weather and are known for having a weather sense.

3- About 656 feet

4- About 574 feet

5- 3.75 miles

6- Considering how nearly everyone and their dog has a dog in Germany, I suspect that since the author did not say two legs but legs, there were probably a few dogs trying to help on the mooring lines.

7- About 10 inches

8- 1917. I'm sure the tree would be indistinguishable from any other tree now, if the tree survived World War 2.

Chapter 4

Attacking London

"What's the real reason you don't go to England more?"

I couldn't count how many times I have been asked that question or some variation of it. And every time, I could only give them the same answer, "the weather."

Once, on vacation, in central Germany [1], I sat at a popular restaurant and just so happened to be next to a table of old, opinionated, military veterans exchanging war stories. Eventually the conversation turned to the German Zeppelin attacks on England. I listened in, quite intently. One of the gentlemen concluded, "The weather today is so good, I bet there will be an attack on London tonight!"

When I left the restaurant, I was greeted by the most wonderful weather. The moon was big and bright. Bright enough you could comfortably read the newspaper with just the moonlight. There were some small clouds, moving along, very high up in the North. Up there the wind was blowing South. To sum it up, it wasn't possible for the weather to be worse for zeppelin missions. In fact, none happened that night. I'm sure the man at the restaurant with the "mind for strategy" would have shaken his head at that.

The best nights for Zeppelin attacks were dark and had little or no wind. Also, the wind doesn't blow the same speed or direction all over Germany. For example, Berlin could have nice weather with no wind, while the North Sea coast wind was pretty stiff. But don't worry. Even if the weather is a little bit bad, the airships and crews still fly. No one regrets bad weather more than the crews on the airships.

The weather predictions for our next mission were actually pretty good. Our mission would begin tomorrow.

In the morning, our orders arrived. "Weather forecast is good. Attack London tonight!"

The Leader of Airships calls on the phone to give more individualized instructions. He's joining the attack on L41.

The commanders of the airships discuss amongst themselves who should take off first, and all other sorts of topics they like to discuss.

The first airship took off at 2 pm.

The crews ready their ships for take off. The engines are tested, so that any problems can be fixed on the hangars. Engine troubles are not allowed during flights.

The "parkour boys", bring the bombs from the munitions depot to the airships, where the bombs are taken from the cart and hung them on the airship. There were all different sizes, even some real big fat boys that needed block and tackle to pick up.

Part of the preparations was opening the giant hanger doors. The ground crew was already holding the ship. At the scheduled time, L111 [2] left the hangar, like an awakened behemoth. Shortly thereafter, the ship was cleared for take off. The engines are started, the ship is thrown into the sky, and everyone shouts, "Let's get England!"

Those staying behind wish us well and wave enthusiastically.

Our course heads west, over Borkum. [3]

The sea became populated with giant birds. The airships assembled from all over, forming a giant squadron. Flying west together was an inspiring sight, one you remember for the rest of your life.

"I wish the English could see this," mused the Watch Officer. "Then they would take their annoying little blimps and run away."

"Each Zeppelin has ….. thousand kilograms of munitions. Making it …. thousand kilograms all together. If they all landed on your head, you'd definitely be missing something.". [4]

Directly behind us, you could see the Dutch islands as little gray stripes. We've passed Ameland and Terschelling. [3]

The day turns to dusk. The sun starts to set. Every now and then, we pass Dutch fishing boats. They fly the flag of Holland. It's painted on the side of their boats to make it easier for the U-boats to see. It's also painted on top of the control room, so the Zeppelins can see it too. The fishermen stand on deck, staring in the air, watching these giant birds fly over. Sometimes they wave to us with their caps or handkerchiefs.

The helmsman reports, "we are about 20 nautical miles from the coast."

Soon land will come into view.

The sun floats over the western horizon like a fat golden ball. Quickly it disappears. The horizon becomes a beautiful reddish gold. Against that glorious backdrop, a well defined, dark streak appears on the horizon. It is the coast.

The under officer on the platform reports, "Land spotted, straight ahead!"

Behind us we can see Yarmouth. [3] That lets us know we're where we're supposed to be.

Now the individual squadrons and ships separate from each other. We have to be careful not to collide with each other in the dark. In front of L111 is L31 and L32. We have to pay special attention to them. If we drove into them broadsides, we all would be nothing but kindling. It is fairly cool up here. It's -7 C [5]. While it's uncomfortably cold for every man on board, no airship man would wish for it to be warmer. Cold air provides much better lift than warm air. [6]. The ship can fly noticeably higher in cold air. For the layman, it is sufficient to state that for about every 3 degrees [7] colder, we can fly about 100 meters [8] higher.

The ship crosses the coast.

"Course South West by South. To London!"

The machine guns on the platform are manned. A sharp eye is kept out for enemy planes, so we can keep those boys under fire.

The two ships in front have disappeared from sight. L111 nears its goal.

The lead machinist comes to the forward gondola and reports that everything is functioning properly on the ship. The engines are maintaining proper rpm's.

"At the moment we are just East of Colchester[3], heading towards London. Around 12:30 AM, we attack! Inform the under officers so they are aware of our position."

The lead machinist makes his rounds to the individual gondolas. Just about everyone on board has some type of map of England that they eagerly use to follow the ship's course.

Directly behind us a long way away is a spotlight, pointing straight up. It appears someone over there heard a Zeppelin and they're looking for it now. In spite of their diligent searching, they don't find one.

"That spot light must be around Winterton[3]," the watch officer assumes. A few minutes later, we picked up a radio transmission from the English, "Zeppelins passed over the wash."

Every possible English radio station retransmits the message. Then, another station reports spotting a zeppelin. Suddenly the radio traffic down there goes wild.

We've been spotted too. But it doesn't matter.

The effect of the appearance of German Zeppelins can be seen now. All the lights are out. Not a tiny ray of light to be seen. Every now and then we'd spot a light, but it went out when we moved towards it.

While bothersome, I wouldn't deny the English this. They've learned to blend in. It's really easy to understand, when you consider the fear these people have. A giant bird comes roaring over you and could drop giant burning eggs at any moment.

"Spotlight ahead!"

Aha! They've heard us and are searching the sky. They don't spot us.

A second spotlight appears behind us. It has more luck. Almost immediately it found us and caught us in its light. In that same moment we saw four flashes of light right next to the spotlight. It quickly repeats. It is one of their batteries firing at us.

"Starboard, 15 degrees West from North! Climb!"

The ship quickly turned to its new heading while climbing. The spotlight loses us. The batteries fire a couple of salvo's in the air. What they're shooting at, I don't think they even know. Maybe they spotted a harmless little cloud, and in their delirious rage, tried to shoot the little bastard.

After down below had quieted down, we returned to our original heading. The small trap that we stumbled into, was added to the map.

It was about 12:15 am. We get to the Thames west of London. The English were trying to blend in as much as possible, but they just made the Thames easier to follow. The Thames, with all its unique oxbows, provided excellent navigational orientation.

"Light rudder, turn East.". The side helmsman receives his orders. The ship is heading for an attack.

The last of our water ballast is dumped, so we can attack from the highest altitude possible.

The Watch Officer quickly inspects the bombing preparations again. Now he opens the bomb bay doors, so the bombs can fall unobstructed, secures everything, and reports to the Commander:

"Everything's ready for bombing."

"All engines, maximum power forward!"

The engineering telegraph rings the orders to the gondolas. At ever increasing speeds, the propellers accelerate. The engines must give their all.

There! – up ahead is a spotlight. First four, then six, then twelve to fifteen, they just keep multiplying, all shining in the sky. Eventually there are so many shining all over the place that you can't count them. They're lighting up a ship that is over the middle of the "city".

It is the L31. The commander, Captain Lieutenant Mathy, and the crew have already completed five successful bombings of London.

The ship is in the middle of exploding shrapnel grenades and hundreds of bullets. The whole ship is surrounded. Small white clouds of explosions are above and below the ship. It's caught by dozens of spotlights. In spite of all this, the ship peacefully holds its course.

Now it's over the middle of the city. Soon we see the explosions of their bombs.

"That's good, quite smooth," said the helmsman. He was right. You could already clearly see the effects of the bombs. Fires have broken out at several locations.

At the same time, L111 was now over the city. Seemingly unnoticed. No spotlights have found and illuminated the ship. They're still too occupied with L31.

The city lay sprawled out right in front of us. There's over a million people down there. The Thamse snakes right through the middle of it all. The traffic on all its bridges is easy to see. Behind us is Hyde Park…three years ago, I strolled peacefully down those avenues. Back then, I would have never dreamed that I would see those trees from the sky, on a fiery night!

"Cleared for bombing!" The order goes to the watch officer.

In that same moment, they spotted us. First it was one spotlight, then a second, and a third; until the entire spotlight club was shining on L111 in unison. They finally let go of L31. It had already done its work and was East of the city.

"Bombs away!"

The first bomb fell. A few seconds later we feel a small jolt in the ship, a hit- in a good spot! Soon thereafter the other bombs hit. Explosions and incendiary go all over the place. Those below are supposed to have a little bit of everything!

We hear the bang of our bombs and the pop of the defensive guns thru the monotonous buzz of the propellors and engine noises. Those defensive guns sound really close.

In the middle of all that there's a ring from the platform. "Several rockets just flew over the ship." comes the report. So, those guys are shooting incendiary rounds! Those are especially discomforting.

"They must be stupid. It will take more than that to take us down," the watch officer tells me, "the best part about it is that all these blessings they're trying to give us will come right back down on their own heads."

Now the ship has gotten away from the city. We keep ourselves to the starboard, so that we can drop the rest of our munitions on the large warehouses and docks on the Thames. Those would be especially nice and worthwhile targets. Once those really start burning, it would be nearly impossible to stop.

The spot lights are still lighting up the ship. They are still shooting at us, like that will do anything.

Now the ship is right over the docks. We drop the last bombs. A moment later we see the effects: bright, billowing fire.

A fairly strong shock goes through the ship. It was from the last fat bomb. A worthy final bomb.

"The bombs are all out," reports the watch officer.

The spot lights gradually released us, because the ship flew over a light cloud layer that was over the mouth of the Thames. Now we can finally take a good look at the city and enjoy the view in peace, and see the results of our efforts! It is a fine, small, fiery lane that "L111" had laid down in the enemy's city.

Right behind L111, L41 and L 51 reach the city. Both approach each other, blinded by the spotlights. Luckily, the commander of L41 notices L51 thru the spotlight's blinding ball of light. Just in the nick of time, the ship pulls hard on the rudder and turns to the starboard. It circles around and resumes their attack.

In northern London, two more ships are at work. We see them in the distance like white cigars drifting in the air. Also in the south, it appears an airship is attacking. We could see the precipitious flashing of exploding bombs.

L111 distances itself more and more from the city. Individual details can no longer be seen, only the bright spotlights indicate where the attack is happening. There must still be airships attacking, since all the spotlights are still at it. Every now and then we see the distorted flash of a bomb hitting.

Under the ship, the mountainous cloud layer has thickened, it's not completely closed off, there are several holes in the cloud cover. We still haven't left England yet. Caution has been ordered, so that we don't get betrayed by some hole in the clouds and a battery pounce on us.

Then suddenly the entire heavens became bright.

Everyone looks to the rear, where the light is emanating.

A fiery, burning ball is high up in the air, but only for a moment. Then it starts to sink, falling, falling, slowly at first then faster and faster.

The cloud layer over the mouth of the Thames is brightly lit up for miles. There's no doubt about it…an airship caught fire and fell, flaming. We could still see it burning on the ground and see the glow of it's bombs going off mix with the light of the fire.

So, they got one of us. Everyone contemplates and wonders, "who was it?"

But we don't have a lot of time to guess or speculate. We just flew into a hole in the cloud cover! Right over spotlights and batteries! It could be…no it must be the battery at Shernetz, that's firing at L111.

Immediately, the ship got caught by the spots' claws of light. They held tight and wouldn't let go. It's a pity we didn't have a couple more fat bombs to smash on their heads. We had no other option than to leave this shooting range post haste. Quickly, to the protective clouds at our rear!

"Hard stern!". The ship hastily turned around and in a few minutes, we're safe in the middle of the clouds. A short while later, the ship is headed North to get away from the battery, then we'll head East to get to the Sea.

We passed the coast at 1:15 AM. We could still see the light from the burning airship in the distance. There are still a few spotlights making streaks across the sky. The main attack appears to be over.

Around 2:30 AM, L111 was at the Nordhinder light-ship [8][3]. Now we radio our report home.

"Location: Nordhinder light-ship. London attacked."

Nothing more. All other details will be reported in writing. Right now, they only want to know if we completed our assignment and if we're having a good trip home.

Soon after, the other airships that participated in the raid started reporting back. All have carried out their orders and are "marching" home.

Only one ship was never heard from again, the L32. We had no doubts. The ship we watched get shot down, that crashed burning into the sea, was L32. [9]

L21 attacked at the same time and got to see L32's destruction as the closest eye-witness. The ship had planted it's bombs in the middle of the city. They were already at the North-Eastern edge of the city when they were hit by the deadly incendiary round. In a few seconds, the ship twinkled like an ocean of fire. The bow tilted towards the Earth as the ship raced to the earth, dragging behind it huge flames. They resembled a strudel. Shortly before hitting the earth, the ship broke in two. The burning wreckage buried the crew and their commander, Over Lieutenant at Sea Peterson. They had already completed many successful missions, now they lay under glowing debris.

Only one of the many ships that attacked were shot down. But what does the loss of one ship mean? The crew, who were our friends and associates, were now gone. Contrasting this with the damage we caused, which the English press never truthfully reported. Somehow our newspapers eventually got word of the damage. Their reports indicated how scared the English were, how many materials we had destroyed, and the impression our attack had left on the English.

They don't have an irrational fear of the Zeppelins. Honestly, they do have to act as if ... and report: "As far as we know, no real damage has been done….an old woman and a suckling child were severely injured…oh yeah! And a bicyclist's inner tube popped!"

But we, those of us who watched from above… we know the truth about the cyclist and his popped inner tube!

*. *. *.

Around 6 AM, L111 wass at the Terschelling light-ship and headed towards the German Bight.

We weren't lazy on our trip home. Once we left the English area, we wrote down our report and everything related to it, so that they could be finished as soon as we landed.

"Attacks on England like this would be so very nice," the Watch Officer started, "if it weren't for all the damned writing!"

He was right. But that didn't help. It still had to be done.

The sun gradually rose in the East, right in front of us. The new day breaks.

Around 8 AM, the first German Island, Borkum, comes into view.

With a good tailwind, we make it home in no time. Below, everything was already for landing. As the ship was skillfully landed and marched into the hangar, we heard "Admiral of the Air" [10] greeting us.

1- In the Kaiser's Germany, central Germany was around Berlin.

2- Rolf Marben's book from 1932, includes a table that claims to include all Naval Zeppelins and does not have numbers higher than L72. But that doesn't necessarily mean there was no L111. Note to the reader, that is the first book I translated, so I will reference it here when called for. It is available on Amazon.com if you search for "Knights of the Air Madison Hasson".

3- Map showing locations mentioned in text:

4- This was released in 1917, the numbers were redacted for publication. But it is a huge flex to let the world know that the war zeppelin's carrying capacity was measured in multiple tonnes (1000 kg or 2200 lbs).

5- About 20 F

6- The physics calculations support this statement completely. Air temperature makes a huge difference. And what's more interesting…the less simplifications you use in the calculations, the more of an effect the temperature difference has. And you don't even need a college degree to know how to do it. You just need to pay attention in highschool chemistry and they'll teach you everything you need to know to calculate changes in lift due to temperature differences.

7- Celsius, about 5 degrees Fahrenheit

8- About 328 feet

9- A boat that functioned as a portable lighthouse.

10- You can read another eye-witness account of L32 going down in Knights of the Air, "The end of L23".

11- Admiral der Luft, you can find it on YouTube and hear some history.

Chapter 5

Battling English Submarines

L13 laid its course to the North-West, heading towards Dogger Bank. The weather was beautiful and calm.

Next to it, to the South-West and to the North-East, were other ships, all searching for the enemy.   Up till then, no one had seen anything. There was a thick, solid cloud layer at about 800 meters [1] altitude. The sea is its usual peaceful calm.

Around 11 AM, L13 was already far away from shore. Nothing to see, but a few Dutch fishing boats that were harmlessly fishing at Dogger Bank.

Then the Commander, Captain Lieutenant Mathy, I've mentioned him earlier, spied something thru his binoculars. It was on the horizon, a short dark stripe, with something sticking up in the middle.

A submarine! What else could it be?!

The ship quickly climbs into the clouds to remain hidden to those below. Eventually the ship comes to rest at the exact altitude that the forward gondola sticks out below the sea of clouds. It's good enough that the submarine doesn't notice us. If it was an enemy submarine, they would dive as soon as they notice us.

That had happened many times before. This time L13 didn't want to let the spotted submarine escape.

Our ship traveled back into the clouds for a bit. We must have been about 5 nautical miles from the submarine. Once more, we carefully steer our ship, with a little bit of ruddervator, below the clouds to see if the submarine is still traveling along the surface or if they've noticed our approach and submerged.

You can imagine our joy, when the submarine we first spotted was not only still there, traveling obliviously, but had turned into more! There were four! The submarines were either not moving or traveling very slowly.

"It must be enemy submarines, wanting to ambush somebody here," supposed Captain-Lieutenant Mathy to his watch officer.

After a while of careful observation, we could tell that it was an English E-class submarine, from its submerged outline. Once we knew one was English, we had no doubts the other three were English vessels also…otherwise they wouldn't be hanging out so peacefully with each other.

Back to the clouds! We must still be over four and a half nautical miles away. In the meantime, the bombs are cleared for use and our direction of attack is double checked. Everything is ready.

Now, according to our calculations, we must be very close to the enemy. So we descend below the clouds again for the attack. Hopefully our enemy hasn't noticed the humm of our engines and disappeared into the sea.

A look below…they're all still there! L13 heads to the closest submarine at maximum power.

At that same moment, they spotted us and started to take some well aimed shots at us. Under and over us, to our left and right, shrapnel and grenades popped. Their salvo's got closer to the ship, but our ship also got closer to them.

If anyone had doubts if these were actually enemy submarines and not our own, the first salvo cleared it up. It was the best identifier. They were definitely English.

One push of a button, and the first bomb drops into the sea.

Have I ever told you how small a submarine looks from 800 meters [1] above it? And to hit such a tiny target with a bomb? That is really hard. To do the best, we purposely dropped two or three bombs in rapid succession and watched them advanced towards the target.

Same thing here.
The first bomb is about 150 meters [2] short. But the second is already in the air, so it is on its way.
"Second bomb is 60 meters short. [3]"
The third or fourth is bound to connect.
"Third bomb is 30 meters short. [4]"
"Horrah!"
The fourth one hit.

We hit the submarine that was shooting at us with everything they had. The bomb hit their deck, just in front of the tower of the submarine. The detonation was so strong, L13 rattled in the air.

The black smock from the exploding bomb dissipated quickly. The bow of the sub sank quickly. The stern stood up with its end in the air. The propeller was still spinning, incredibly fast. In a few seconds, the entire ship sank into the deep.

As soon as we watched our bomb hit, L13 headed back to the clouds. We want to take out a second submarine.

Quickly, we popped out of the clouds again, but there was nothing to be seen anywhere. The submarines dove down and disappeared once they realized what happened to the first ship.

There's nothing left of the sunk submarine, and no other submarines to find. Just a large oil slick where the submarine sank.

For a long while, L13 patrolled the area. Maybe one of those submarines would dare to surface. But nothing dares anything.

The altitude helmsman reports to the commander, "The ship's rear is getting heavier. We're probably leaking gas from a rear cell."

At the same time, the sailwright came out of the walk way, reported that cell five was floating up. He had noticed it walking through the ship and came to the front to report his observations to the commander; also cell six had a slow leak and was down 10% already.

The ship has been hit a few times during its duel with the submarines. It was no surprise. Missing a target as big as us from 800 meters [1] took talent.

Our next order of business was to drop ballast water to keep the ship from sinking.

The commander walks through the entire ship. He wanted to see the damage himself.

Of all the!...cell five is nearly empty. It must have a huge hole on top for the gas to flow out so quickly. Cell six is about half full.

Due to the loss of lift, the aft of the ship hung down. We continued flying forward with a ten degree incline.

The commander orders the helmsman, "Heading South East, we're going home.".

"Have the watch officer radio the following: Position - North of Dogger Bank. English submarine destroyed. Heading in due to loss of gas."

The holes must be pretty big. The cells emptied so quickly, there was no way it could be repaired mid-flight. The rear of the ship keeps sinking lower. All available water ballast had been dumped and the ship still wouldn't fly level. We had to work against it.

Everything in the rear that wasn't nailed or riveted down was hauled to the front of the ship: oil cans, tools from the rear gondola, machineguns, ammunition, bombs, leather suites, coats, lines and ropes, etc... Pretty much anything that can be is "trimmed" to the bow. Gradually, it works and the ship levels out. The ship was spotlessly trimmed up until we landed and the ship was weighed off.

After a smooth landing, the ship is brought into the hangar. That's when we found out cell five had two large holes in it. It was a direct hit that went straight through the whole ship. Cell six had been hit by a couple of decent sized explosions, and the rest of the ship had been damaged by many small explosions. Just below the rudder there was another direct hit. Twenty centimeters [5] higher and the ship would have been unable to steer.

The repairs were quick. It hadn't even been two hours by the time all damage was repaired. We could report that the ship was cleared for operations again.

The ship and crew had the honor of being the first to ever sink an enemy submarine from the air. We celebrated with our shipmates, but there was a hint of jealousy in their jubilation.

.*.*.*.

1- About 2625 ft, a half mile.

2- About 164 yards.

3- About 197 feet.

4- About 98 feet

5- 8 inches

Chapter 6

Traveling in a Storm!

We had another successful attack on England under our belts. This time we covered the industrial facilities of the cities in the middle of England extensively with bombs.

At midnight, L111 was high over Flamborough Head, heading home going East at max speed.

We inspected every part of the ship. God be praised, no where in the ship was there any damage to be found, in spite of their spirited defense.

"Aft, starboard side, there appears to be strong weather," I commented to the helmsman standing next to me. My estimate put the storm somewhere over Amsterdam. It looked quite severe. After a while I ordered, "Take bearings from time to time, so we can see if it comes any closer. I worry it's going to come North to the Dutch coast, but not go out to sea. For now, maintain our course."

The Helmsman measured the storm that was clearly growing in the expanse. Meaning, he determined the magnetic bearing that the storm had in relation to the ship. The observations revealed the storm was actually moving North-North-East at a very fast clip.

"Maybe we could still avoid it. If not, let's go a little further north, back out to sea, and go around the storm's northern end."

The sky above us was still crystal clear. To the East and South thick clouds were moving towards us. That bode ill.

The friendly west wind, that until now had been pushing us powerfully, grew weaker and weaker. The ship wasn't making much progress traveling over the sea. It was very clear. The lights of the vehicles we occasionally saw on the water were moving very slowly out of view.

"The storm is making a lot of static," Lieutenant at Sea Von Schiller reported as he came out of the forward gondola radio booth. "If it gets any stronger, we'll have to shut off the radio. The radioman can barely receive anything as it is."

"I can imagine. Go check, see how things are to our South East. The worst part is that stuff keeps moving closer."

"Starboard side ahead, lightning just struck," reported the under-officer at the steering wheel, "But it was pretty far away."

For a while we strain our eyes looking out ahead. Yep, there's a storm there too. That must be just north of Terschelling.

"Two strikes aft, East side of North."

We're just before Dogger Bank and want to veer slightly North, to escape the storm.

I had the head engineer come to the lead gondola, explain to him our situation and asked him how many hours of fuel we still had on board.

"We're good for eleven hours," he responded.

That was still a considerable mass. But, in the end, it wasn't enough to go around the storm. We hadn't planned for a storm, nor could we handle headwinds.

The lightning strike became more frequent. We could see the individual veins of lightning.

A signal whistle sounded. The radio booth was calling.

The Helmsman hurried to the speaking pipe.
"Warning"

"Radio has been shut down due to the increased interference from the storm."

L111 was currently in the middle of Dogger Bank.

The East, South-East, and South were all full of storms. It was a large front full of heavy clouds blocking our way home.

What to do?

The wind had changed to out of the South East and got noticeably stronger. The ship did not make much more progress. The lights below us that showed the locations of fishing boats weren't moving.

"Course North!" I ordered the helmsman.

I thought we would try to go North to go around the storm bank.

The ship turned North and suddenly made good time since the wind was now at our backs.

We couldn't travel too far North, we have to travel against the wind at some point in time. That will use a lot of fuel.

"I don't think we're going to make it through this way," the watch officer stated after 15 minutes, "that crap keeps coming closer."

I ordered the helmsman, "Starboard, ten degrees, course South."

"We'll see if we can get around it going South."

Clouds are zipping by below us. First a few, then more and more. Things got gustier. The ship started to buck, meaning the nose would go up, then drop down.

"We're not going to get away from the storm heading south either!"

After an hour it was clear there was nowhere to turn to get away from the storm. Our best chance was to go West, but that was the way to England and we had just been there. We'd rather not.

"It's starting to rain," chimed the lookout on the platform.

The clouds, which looked like thick mountains, were getting taller and taller. The ship had to pass through those mountains. Those mountains contained things that would shake the ship like a rag doll: hail, snow, rain, and especially gales.

Soon we're 100 meters [1] above the altitude I've ordered, then we're suddenly 200 meters [2] below.

The lightning strikes became more frequent and stronger.

There's no escaping it now.

So- Through!

Hopefully the storm bank is not as deep as it is wide, so we can get through it quickly.

A piercing crack sounded! Instantly it was as bright as noon-day. The entire gondola was lit up.

Lightning just shot down to the Earth right next to the ship. [3]

The sailrite, who was on lookout on the platform, reported, "the rounds for the machine gun were burning."

That phrase seemed a bit Spanish to me. [4]. Therefore, I sent the watch officer to the lookout platform to see what was really going on. He was also supposed to use the speaking tube to tell me where the individual lightning strikes were coming down. This way we could steer the main body of the ship to avoid the lightning.

When he got up top, his eyes opened wide with surprise. Everything was glowing. And sitting right in the middle of it all was our small sailwright, soaked to the bone, with a literal halo over his head.

It was what's usually called "Saint Elmo's Fire". [5]

The large aluminum frame of the ship had become so charged from the storm that it was radiating electricity from every point, sharp angle, and edge. That was also the cause of our "Burning rounds".

Lower on the ship, they could start to see the signs of the static electricity. The guy wires and cables all started to glow blue-violet. It was a beautiful picture. The situation the ship was in, no one could have known how beautiful and enjoyable the experience was except God, or those who were there.

"Ahead, port side, the lightning is really strong," the watch officer reported from the platform.

So, we needed to veer starboard. Our course is generally South East. We must get through!

It is raining buckets. The ship is sopping wet.

Now we're right in the middle of it. It couldn't get any stronger. You couldn't even tell where the lightning was coming down. As soon as you saw one bolt strike, a stronger brighter bolt would strike. Right and left, over and under the ship, the giant bolts swished into the water, from cloud to cloud.

It is now constantly as bright as day in the gondola. Over and next to the ship were giant cloud mountains. On the edges, the gusts are racing around especially strong.

About every one to two seconds, lightning strikes. In the middle of all the magic, 3000 meters in the air [6], was an airship that couldn't tolerate even the tiniest bit of fire…I can imagine much more pleasant situations I would rather be in!

Actually it was quite a bit like attacking London; but in the end, you could defend yourself with all types of maneuvers. And you could settle your nerves, by the sounds of our own bombs falling and exploding. But here…there was no equivalent. The ship was simply a toy ball for the elements. You only needed one thing in this situation, luck!

That's how things had been going for the past two and three quarter hours. Eventually the storm weakened, and shortly thereafter stopped. Finally, we could breath easy and gain strength.

Twilight must have started around 4 pm. We awaited nothing more eagerly than daylight. When you could see, everything was different.

The helmsman said, "Now it seems like it is actually stopping.". At the same time the watch officer up top reported, "the rain is letting up, we can see stars every now and then."

We have reached the backside of the storm front. Strong gusts shook the ship back and forth, but we were out of real danger.

Things went well. We made it through!

I let the watch officer come back down, and relieve the under officer up top and the helmsman.

"That could have easily gone bad," one said dryly to the others.

"Yes. The man must have luck."

The wind also starts to slack off. South-East in the German Bay, the weather is beautiful.

The sky starts to turn red and golden-yellow, and a few minutes later the red ball of sun slowly rises over the horizon.

It does us all good, ship and crew.

The ship, which was totally wet, started to dry out. It was clear the ship was losing moisture, the ship was becoming lighter, meaning we gained altitudes.

The crew, who were not on watch, made themselves cozy and started to eat breakfast. They had all received a piece of sausage and a roll in their provisions, and were enjoying it immensely. They finished it off with a mighty swig of their flasks. But it wasn't Schnapps or anything like that, because any consumption of alcohol was strictly forbidden. [7]. It's coffee, or as we more correctly say, "it's deceptively like coffee.". We carry it along in a thermos. [8] We also had a small electric cooker, to boil water for tea or cook some eggs.

Epicurean delights weren't ever expected on the menu, not because we couldn't get them, but because it gets heavy very quickly when you provide world class provisions for everyone on board. On an airship, every kilo, every gram of unnecessary weight must be left at home.

The most uncomfortable rule is the no smoking rule. But you can eventually get used to that too.

A lady once told me, "I would have loved to give you and your crew cigars or cigarettes as a sign of my affections. But you're not allowed to smoke on airships and have probably all given up smoking because of it."

Nonsense! We aren't always traveling around in the air! That would eventually become boring. After we've landed safely and the ship is properly stored in its hangar, we love to go to the mess hall, light up a couple cigars, and tell stories. [9] [10]

The sun is already high in the sky. The watch officer and sailrite both start to dry out a bit.

Helgoland comes into our view, up ahead. We had expected to see it earlier, but the storm had such a strong head wind that for three hours we ran in place.

Two hours later, we arrived at our home port. The ship lands and was brought into the hangar.

We had become much richer due to our interesting and varied experiences. The ship passed a very thorough inspection with flying colors, and no other mission bonded the crew to the ship better than this mission through the thunder and lightning of a stormfront.

1- About 328 feet.

2- About 656 feet.

3- This could sound like hyperbole in English, but in the original German, the author quite literally meant directly next to the ship.

4- This idiom is similar to "it's Greek to me". One of the emperor's of the Holy Roman Empire was raised in Spain. The other nobles had trouble understanding him at times. This idiom indicates that he understood the words being said, but didn't understand what the words were supposed to mean all together or what the sailrite was trying to say.

5- Saint Elmo's Fire happens to ships during storms. The ship builds up enough static electrical charge that the top of the mast starts to glow. If you look up "Saint Elmo's fire weather images" online you will be able to find pictures and see the purple color described in this chapter. If you leave off "fire weather images" you may get a movie or song from the 1980's.

6- About 10,000 ft

7- In "Knights of the Air", the chapter titled "The Sailrite's Tales", the alcohol policy of the Imperial German Navy Airships is explained in great detail. There were times alcohol was consumed, but only on high altitude flights. This flight may have been before high altitude provisions and policies had been implemented.

8- In German a thermos is a thermos-flask, so frequently they leave off the thermos and call it a flask.

9- The word in the original German text was "Bratjen". I couldn't find a translation, so I used my best guess. It could just as easily be translated as jokes.

10- By no means is my translation of this German's positive comments about smoking to be taken as my agreement or support of smoking. I am personally morally opposed to smoking, but I feel I can accurately translate this author's text and not condone smoking at the same time. Accurate translation of this historical text helps the reader understand the culture of the Kaiser's Germany. When I lived in Germany from 2002-2004, smoking was still very common. If you were in a public location, you would smell tobacco smoke in some form. It was still very much part of their culture. I do not know if that has changed.

Chapter 7

Encounter with English Sea Forces

It started to brighten up. The airships were already out at sea, reconnoitering the North Sea. L111 was headed west-north-west and just as the morning twilight appeared, we were just north of the Isle of Borkum. Next to us, a bit to the south, was L41. A bit to the North was L66, our old ship. It brought us such joy to see our old ship for the first time after we had been reassigned.

That's how our reconnaissance went.

About 7 in the morning we received a radio message from L19, "Several smoke clouds spotted north of Terschelling."

L19 was on the southern edge of the reconnaissance zone. It was one of the last missions the ship flew.

A few minutes later, L19 sent another message, amending the first, that the smoke clouds were emanating from enemy ships.

"Without a doubt, they are up to something down there!"

We maintain our course. Maybe there are also enemy forces in the North too.

L19 quickly approached the enemy. They got close enough that they could tell exactly what kinds of ships they had.

It is three large steamers, with large wooden construction on top. They followed each other in a line, but with a lot of separation in the gaps.

Airplane motherships! They're airplane motherships! [1] No doubt the enemy intends to do an air raid on the German North Sea coast!

The motherships were carefully protected and had many lookouts. In front of them were two small cruisers, warships lined up along either side of the motherships, including armored cruisers, and two small cruisers at the far end. They were so far away it was hard to see them.

To defend against German submarines every ship was accompanied by a torpedo-destroyer. They had made a pretty good wreath of warships to protect their precious cargo.

The whole swarm was headed east, to the German Bight.

After L19 broadcasted the specifics about their forces and heading, other airships headed that way to feel out the enemy forces.

One airship after another volunteered to join in on the effort.

This made our opponent very uneasy. They probably wanted to get to German waters unnoticed before releasing their planes.

L111 went south-west of the enemy to get a feel for the situation.

The beautiful weather offered very good, long distance visibility. The East was foggy.

"The small cruiser and its flotilla are changing course, heading North," the Helmsman reported.

The enemy is splitting their forces. Probably hoping to misdirect the airship's attention or to cut off the path of some of the airships in the west.

After he took his position at the tail end of their forces, L111 followed behind him.

Up at altitude there was a lot of west wind. So we quickly approached the enemy vessels. Another group split off heading to the south-east.

As soon as we got too close to the enemy, we stopped, to avoid being in their shooting range for too long.

Suddenly, we could see the outline of many ships in the fog.

The first one we recognized was an English heavy cruiser. It was headed north-west, straight towards us.

We had just barely seen it when we heard the thunderous roar of their first shots.

"Hard starboard, reverse course," I ordered the helmsman.

We were in quite an uncomfortable situation. The strong wind helped us approach the English quickly, but now it was hindering our retreat.

"All engines maximum power.". The engineering under officers also saw what was going on. They saw the white and black puffs of explosions in the air. We all knew the only thing that could help us was our higher speed.

So open the valves and let the engines run as fast as they want.

The watch officer called to me, "They're shooting at us with heavy artillery!". Rightly so, all their heavy cannons were flashing. Every now and then their mid-sized artillery would fire from the casemates. [2]

The explosion clouds kept getting closer to the ship.

"We are slowly out running them," reported the helmsman, who was constantly measuring the distance.

Both light cruisers headed off to the south-west. Apparently, they wanted to catch us in the middle and cut us off if we headed south-west.

"Higher!"

The altitude man quickly took the ship up.

Enemy fire helped us know how right our timing was. It exploded right under the ship and gave it a light rattle.

The engineering telegraph rang.
"What's up?" I asked the helmsman.
"The starboard rear motor stopped."

Fiddlesticks, [3] I knew we were missing something! Naturally it happens now, when we need all engines.

"Head south!"

We must ensure that we get to the Dutch coast before both light cruisers.

Until we get over the coast, we're just a giant floating target. We're showing them our entire port side. On the other hand, sudden course changes disrupts their aim [4], additionally "every cannonball doesn't accurately hit its target".

Pretty soon we're right on them. We were about five nautical miles away. Then the heavy cruiser and light cruisers stop firing at us....why?

We are between the enemy and neutral ground. Every new salvo would land on Holland, neutral ground.

The ship turned east and used enough speed to maintain position.

No one could touch us here.

Our enemy saw this too. After a while they headed north and rejoined the rest of the strike force.

L111 slowly followed them.

In the mean time, other airships had harassed the rest of the fleet with their bombs, especially the airplane mothership. In response to this, they hastily towed in the aircraft they had already deployed in the water back on board. Mostly because they were certain the German fleet was coming, and they had no desire to engage it.

They turned west and headed back to their home port at top speed. They were persecuted by the airships for quite a bit longer, till the afternoon. Due to the impending darkness, contact was broken off.

The airships went into their hangars. A few had been hit. It damaged their cells, but they all made it happily home.

The well staged plane attack on the German North Sea coast, that was supposed to be their answer to German Zeppelin raids, was horrifically derailed.

And whose fault was that? The "damned zeppelins".

1- Airplane motherships were the first type of aircraft carriers.

2- Casemates are guns built into the side of the ship.

3- In the original German he used the word "Thunderstorms" like we would use the word fiddlesticks.

4- The original German used the phrase peaceful shooting, using peaceful meaning as in calm, unrushed, smooth and steady aim.

Chapter 8

Scouting with the Fleet

About 8 pm the commander left the office of the Leader of Airships, after receiving orders for their next reconnaissance mission.

Eleven o'clock that night, the zeppelin field was bustling with life. All hands were on deck because every zeppelin was taking off.

L111 was supposed to be ready for take off at midnight.

The ship was packed full of gasoline so we could spend the most time possible in the air.

I was required to bring an officer on board, so he could be informed. He had never been on an airship before.

We both went to the small, but nice and comfortable mess hall, to pass the time quickly till dinner, when we could eat and strengthen ourselves for the mission.

"I am so extremely happy for my first ride on a zeppelin!" He told me. "When do we start?"

"You must not speak that way. It will take some getting used to," I replied. "Airships don't 'fly', they 'float'. Float like every other ship. They also don't 'start'. They 'set off'. Whereas airplanes 'start' and 'fly'."

Naturally, airships have their own terminology, much of which was derived from nautical terminology. It's also no surprise. After all, what is more related to an airship other than, for example, a submarine?

My guest asked a whole lot of questions. Naturally they were all the same questions that any layman would ask. Their interests tend to be drawn towards similar things.

"Be aware, I get dizzy easily. For example, I couldn't walk on the catwalk under the roof of the hangar without convulsively holding tight to the railing."

"In the hangar?...I believe it. That's also something totally different. When floating in an airship you haven't the faintest feeling of dizziness, not at take off, not in the air, not even when you stick yourself as far out the gondola window as possible. That's because you feel completely disconnected from the Earth, unlike when you're standing on the catwalk in the hangar."

"What about seasickness?"

"Seasickness?...none has ever gotten seasick on an airship. That is probably because all the movement you feel on the ship feel so smooth and small, even when traveling in very gusty weather. The painful, sudden, jarring motions of the sea are simply not there. If you want to get seasick, I recommend you ride in a tethered balloon. You could do that here too. My first time in a tethered balloon I was with my friend, Over lieutenant at Sea Petersen, who recently was shot down while commanding L32 in a bombing run over London. The tethered balloons had just arrived. We wanted to learn about them too. There was a lot of wind. After a quarter of an hour, we were both sick. No one was willing to admit it though. But our facial expressions revealed the truth. We grew pale and paler. There were plenty of people waiting to get their turn in the basket, so we didn't mind coming down. We didn't mind that we were interrupted by a telephone call informing us they wanted the balloon taken down.

Had we telephoned that we were not feeling well, they would have left us up there until we produced physical proof of our condition. When we first got down we both stood there, "It was only 15 minutes before I got seasick.". There are people who can handle that kind of stuff better, but as I said, you feel absolutely nothing of the sort in an airship!"

I checked the clock. It was already 11:40 pm.
"We've gotta hurry!"

"If you could excuse me for a moment, I can follow you to the hangar in five minutes."

"We don't have any time to lose! Besides, you can do whatever it is on the ship."

"What!?!?!". He questioned, stunned and disbelieving.

"Naturally. Somewhere quiet in the rear of the ship, but – without any water for washing!"

We walked quickly to the hangar. One of the two ships that were supposed to take off before L111, were already in the air. You could hear it drone.

As we walked into the brightly lit hangar, my guest stood in place stunned.

"That thing's enormous!"

"Not true," I countered. "It is really small compared to the new, really big ones."

"This is going to be our last trip on L111. In a few days, we'll hand it over to a new crew and get ourselves a new bird."

"Who gets the crew?"

"They go with me, naturally. The entire crew gets a few weeks vacation, then we meet up at the ordered time in Friedrichshafen."

"Ship is clear for departure," reported the watch officer, who was walking towards us.

I tow my guest into the gondola. There he received a fur coat and felt boots, so he wouldn't freeze. Then he was placed on a nice little wicker chair so he could look out the window and watch the entire maneuver.

The battery whistle sounded.

At exactly 12 midnight L111 left the hangar. After a few minutes the ship is in the air and heading towards the coast.

Every commander has received a sealed envelope containing summaries of the immediate orders for their ships.

Now, after a successful take off, it can be opened.

"The fleet is underway, the airships shall scout."

L111 shall go to Moray Firth [1] and then head North.

An impeccable assignment. It could be a beautiful, long trip.

It is still dark night. We travel 500 meters high [2], as we see bright beams of light under us in the water. A closer look reveals them to be ships. That is the fourth squadron. All ships, completely dark, there is no glint of light to be seen from outside the ship. Even though they seem to be going at high speed, L111 clearly overtakes them.

The third, second, and first squadrons get passed. They're going the same course we are. The fleet's flagship is told who we are. Later heavy and light cruisers come into view. They are supposed to protect the capital ship. The torpedo boat flotillas that are prowling about the water at high speed have the same mission.

"That looks glorious," my guest told me.

"Yes. Maybe we'll meet up with the fleet during the day, when you can see it really well."

About 6 AM was first light. It was a glorious day. Hopefully, we could find something of the enemy, maybe very far away, so that we could bestow upon them a second loss at Jutland [3].

L111 was pretty far out, when the sun rose. Minute by minute, our visibility increased. Far and wide there was nothing suspicious to see.

To our starboard and port an airship appeared perpendicular to us, heading northwest.

To our right was L71, left L63.

"Would you like a tour of the ship?" I turned to my guest.

"Yes, naturally, very much."

We took off our thick felt boots, so that we could climb up and down the ladders better.

I turned over command to the watch officer: "Heading North West Half North. 1200 meter [4] altitude. All engines high speed. If anything goes wrong, telephone me immediately. I'm going through the ship."

We climbed the ladder to the gangway. It wasn't very pleasant. I would recommend wearing a cap like my companion's due to the wind from our speed.

"It's enormously drafty here!" [5]

"Yes....You must remember, we measure our speed through the air in kilometers." [6]

We followed the gangway to the rear. I explained to him the set up of the ship, gasoline, and water distribution.

In the middle of the gangway, the so-called teamroom, the off duty watch were for the most part sprawled out and sleeping. Naturally, downbeds or anything like it weren't on board. They all had a wool blanket, wrapped themselves in it and immediately turned into sleeping mush at the droning sound of the propeller.

We got to the rear gondola on our tour, and climbed down into it.

It's noisy as hell in there. Communication is best done with hand signs and waving. If you want to say anything to anyone, you have to shout into their ear. But there isn't much to talk about. Every under officer watches his engine, feels it, listens to it from time to time, and he receives his commands through the engineering telegraph. Exactly like on a sea ship.

We climbed back up into the gangway.

"That is a horrendous noise down there. How long are the men on watch?"

"Normally they are relieved after four hours. But when something is going on, like "Planes Sighted!", when the machine guns need to be manned, then it could be longer."

"No...I could stand that noise for so long."

"It's all about what you're used to. In the end, this airship isn't supposed to be a nursing home, right?"

We head back to the front.
We climbed the stairs up to the platform. It's actually really hard work. They're very steep, and when the ship is at high altitude, where the air is noticeably thinner, no one wants to climb those stairs.
But anyone that does is greatly rewarded.
Up top it is very peaceful. The noise of the propellers and engines are completely dampened by the hull. There is a windscreen set up in front of the platform, behind it you can sit at the lookout post and peacefully observe everything. Next to it is the speaking pipe, with which you can communicate with the lead gondola.
It was also wonderfully warm up top. The sun beamed down unobstructed on the ship, and the gas below warmed up. The spot was made for sleeping.

After a short time we climbed back down and arrived directly into the gangway, when I noticed the noise of the engines and noticed they were going "slow".

At the same time, the nose of the ship dipped down, and continued downward.

"What might be happening?"

I ran to the forward gondola and climbed down into it and looked out.

Below us was a seaplane resting in the water. It had an iron cross on it, so it was our own.

L111 descends and slowly approaches.

I took the speaking cone, usually called a whispering bag, and as we got close enough I asked, "is anything wrong?"

"Water plane 606 from Helgoland. Lead Lieutenant at Sea X. I have engine trouble."

"Do you mind floating around for a little bit longer?". I studied the situation more.

"Yes, the sea is really calm right now."

"Helgoland is being informed," I called back to him.

"Thanks a bunch!"

The engines receive their orders to press on, and the ship quickly climbs back up to altitude.

"Radio Helgoland! 10 AM, 120 nautical miles NW1/2N from Helgoland, found plane 606 engine breakdown, request assistance."

A torpedo boat received the order to go and tow him in.

To the starboard ahead, we see smoke clouds. We approach quickly. We can recognize their national colors from far away, since they had painted it on the side of the ship. They were Norwegian.

It was fully loaded, resting deep in the water, heading towards Firth of Forth.

We immediately radio our report, and soon thereafter a torpedo boat comes to inspect.

It's a scene we frequently see. A lot of their merchant marines get caught this way. Our torpedo boat inspects their ship, contraband gets found, the ship is taken prize and a boarding crew takes the ship back to a German port.

It was getting to be around noon. We should see the English coast soon.

Ten minutes later, the coast around Aberdeen comes into view. The city and its large breakwaters [7] are widely known.

But there are no enemy forces to be seen, anywhere.

L111 heads north. In Moray Firth [1] we saw a couple of small ships, they looked like small coastal steamboats. Then we headed North again.

Until the evening twilight we didn't see anything noteworthy. We did see a few steam boats and reported them.

In the northern North Sea is a full cloudbank, around 1200 - 1800 meters [8], that could come in handy.

After it got dark and we couldn't see the ships anymore, we went over the clouds. That way no one could see us and we wouldn't receive any surprise fire from ships.

In total peace, we eat our dinner. Everyone not on watch makes their way into the ship, because it was getting cold. Everyone finds a spot and turns to mush.

.*.*.*.

On the next morning, at light, L111 is back to the same spot we were when darkness set in.

Today we shall reconnoiter from England to the Norwegian Coast.

We didn't see any enemy forces, and all the other airships are finding the same thing as L111.

"We're not seeing a lot of steamers," my guest turned to me. He had spent all night writing in a wicker lounge chair.

"Yes, they're getting fewer and fewer. A large part of them are laying on the ocean floor, the rest are scared that they'll somehow get caught in the fingers of our German U-boats."

It was three pm when the Norwegian Coast came into view. In this clear weather, the bald, steep mountains can be seen from very far away. The snow capped peaks offer a wonderful view.

"In front of us is the island Utsira [9]," I tell the helmsman, "We want to approach no closer than 5 nautical miles, then head South staying close to the coast."

We would have liked to approach closer so we could get a clearer view of the land, and I'm sure the Norwegians would have enjoyed seeing an airship up close, but that's not allowed. The airboarder (3 nautical miles from the coast) must be respected. [10]

As we marched south, we suddenly saw signals in the sky.

"One of our submarines!"

"U186," At least that's what was on the identifying flag on the mast.

It was heading home. It's been out to see for many weeks waging war on enemy commerce. It had sunk quite a lot of enemy steamers.

Maybe we can guide another prize into its arms.

With a wind out of the North, our trip home went by quickly.

We took a small detour to search Skagerack [11], but there was nothing to see there either.

About 5 pm ahead on the starboard side, we saw large smoke clouds.

It was the High Seas Fleet.

We plan on getting a closer look.

Pretty soon, we caught up with them and flew by every squadron, which were traveling single file.

"Fifty meters [12]," I ordered the altitude helmsman.

L111 traveled very low and close to the line of ships. Everyone waved at us, and binoculars were used to see if they recognized anyone.

Then we heard a long whistle followed by a short whistle coming from the battery [13] whistle. Turn the front to starboard.

Everything appeared to stand still as we passed the fleet flag ship.

That's when they honored us. We clearly heard the signal of the commander, "front to port."

On top of the command bridge is the fleet officer, waving at us.

Another long whistle, "Get some rest!"

We pass the fleet and head towards Helgoland.

After a long time we could still see the fleet. Against the setting sun they look like dark points jutting up, each with a small flag of smoke.

Such a wonderful, exciting view! The whole German high seas fleet on the North Sea, looking for their enemies to confront and defeat them! But there are no enemy ships outside.

We reached Helgoland by dark.

We radioed, "Landing at 9:15pm"

"Isn't it really hard to land at night?" My guest asked.

"It's exactly the same as landing during the day."

L111 slowly approaches for the landing. She descends even slower to the ground.

"How high?"

"50 meters," [12] answered the signal mate on the ruddervator.

"Steer downward, all engines stop!"

Shortly thereafter the ground crew grabbed the mooring lines. The ship landed on the field. It is quickly brought into the hangar. We must hurry because we can already hear another ship coming in for a landing.

It is L71, who was also scouting, has returned, and wants to land.

The ship is fastened down while in the hangar.

Immediately the crew and maintenance personnel prepare to make the ship clear for travel.

Gas[14] is refilled, gasoline is refilled, the engines are inspected, cleaned, and more.

In one hour the ship is ready to go.

My guest and I go to the chow hall together to enjoy some dinner. Then we enjoy a cigarette! The long abstinence from this pleasure drives you to enjoy one as quickly as possible. And the first glass of beer after a mission, tastes amazing! [15]

. * . * . * .

At 8 AM the next morning, we handed over the ship to the new crew. Since everything was prepared ahead of time, the transfer went quickly. It was all taken care of by noon. With well wishes and admonitions, the crew are sent on leave to visit home.

"See you again in fourteen days in Friedrichshafen when we pick up L130!"

1-Moray Firth circled in blue .

2- About 1640 ft

3- The Battle of Jutland is called The Battle of Skagerrak in German accounts.

4- About 3937 ft

5- In Germany, they have a very strong superstitious belief against drafts. Any moving air is considered incredibly deleterious to your health, even if it is over 100 degrees outside. So even if it was hot in Germany, they still wouldn't use air conditioning or fans, because of their superstition against moving air.

6- Yes, I know that speed is technically measured distance/time, but for some reason this was the German parlance at the time. I guess he means they go so fast, you wouldn't use meters as the unit of distance to measure their speed.

7- A breakwater is a structure built offshore to break the waves before they get to land

8- About 3,937 - 5,906 ft

9-Utsira on a map:

A map of Utsira

10- This is most likely an old international agreement that has been superseded by other such agreements in our day.

11- Skagerrak shown on a map

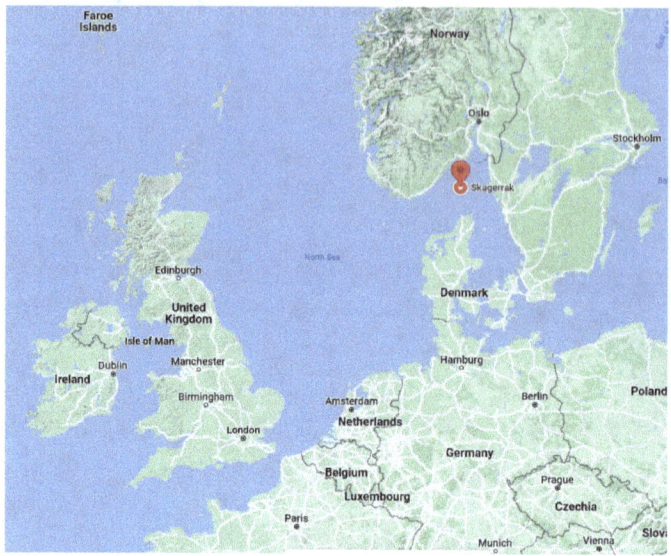

12- About 164 ft.

13- A battery in this case is a place where ammunition is stored.

14- Lift gas, hydrogen. Not gasoline.

15- Beer is very ubiquitous in German culture. I don't drink it, but I know most Germans love it. They take it quite seriously. One of my favorite bumper stickers I ever saw was related to this: "When the environmentalists finally get what they want and all gas stations are shut down, only then will you realize that nobody else sells beer at night."

L130

Our pleasant time in Friedrichshafen [1] came to an end.

Our free days were used to take excursions into the area surrounding Lake Constance.[1] Many old acquaintances and many old haunts [2] were visited.

The day of the first test flight had arrived.

We went to the shipyard on a muggy, hot summer day. On the way, we met Count Zeppelin, who was also there, and accompanied him. Naturally, we spoke about the new ship, the first of a larger type, and smiling he said in the course of our conversation:

"I was already thinking of building ships this big back in 1900, but if I had told people back then, they would have declared me completely crazy."

The times changed quickly. Today, the Z1 could be a lifeboat for the L130.

We went into the hangar to see the ship. The giant was laying in the massive hangar. The finishing touches were being completed.

Around noon, the ship was brought out and rose up for its first trip. The moment an airship takes off for the first time is much like the launching of a sea ship.

For both it is the first time they are handed over to their native elements.

The trip had no issues, and we could leave immediately.

The old count had us over that evening. We will never forget that hour. Naturally, we spoke mostly of airships and their uses, but we also spoke of his youth, his experiences in the war of 1870, and above all, his research and building of airships. He told the stories in such a lovely and refreshing way, that you really couldn't tell that he was eighty-seven years old.

At about 1 pm, we went our separate ways.

We were especially happy, when he shared that he wanted to ride with us on the ship's training flight to its future base.

8 AM the next morning, the weather was wonderfully clear, we took the ship up and headed towards the Swabian Alps.

Graf Zeppelin carefully observed the path of L130. He knew the area below, his homeland, like the back of his hand, and recognised every city and village from above without using a map.
"There in 190*, I got the old Z* stuck in a pear tree.[3] The nose had been badly damaged, but we made it work by patching the hole, so the ship made it back to Friedrichshafen undamaged."

I told him about my trip with L66 into the forest.

"So we're tribulation brothers," he said.

"Yes, your excellence. But I picked a spot close to the hanger to bang my ship on again."

Behind Heidelberg, we saw a D-train, going full speed in the direction of Frankfurt. We went very low. Below us, everyone on the train waved and watched us race against the train. Count Excellence found a lot of pleasure in that. We playfully passed up the train, and soon the only thing we could see of the train were the white steam clouds. Ten years ago, the race would have turned out differently.

We were already in Frankfurt when the train arrived at its scheduled time in Darmstadt [4].

At 3 pm, L130 landed. Count Zeppelin stepped out of the gondola and greeted everyone present.

Naturally all hands are on deck when a new airship arrives from the factory. They want to see it. Today, there was even more interest, because the count himself was on board.

In the evening, the Count joined us in the chow hall. He didn't leave until 1 AM. He couldn't accept our invitation to stay another day. He had business in Stuttgart and was already on the train heading back at 5 AM.

There was a lot to do on the ship for the next few days. The bad weather was pretty convenient at the time. It was impossible to fly in.

Even on the days the ship doesn't fly, there is a lot to do. Such big ships require maintenance if you want to keep them ready to go. When the work is finished, we play sports. There is no better place to do that than in an airship hangar. Soccer and Rounders [5] were enthusiastically played. Sometimes we all went on an excursion. Even gardening was done. We never got bored, even on the days we couldn't fly. In spite of this, everyone was happy when after several days of bad weather, it was good flying weather again.

When L130 was heading home after its first major flight, there was a thin curtain of mist over the coast.

"We're definitely going to have fog," I told the Watch Officer. "If we're lucky, maybe we'll land just before the fog gets heavy."
But apparently, the fog got really dense really fast. Over land it was already really thick on the ground. Anyways we could barely recognize anything. Every now and then we spotted a house, or a street, or a small stream. But the further inland we went, the thicker and more impenetrable the fog was.

Fog is the biggest and most horrible enemy of sailors. That's an old saying. But this saying is even more poignant to the airship man. A ship can usually drop anchor when it's too foggy. An airship is stuck in the air and can't land. Because no one can tell exactly where the airship is. Navigation becomes especially more difficult, because usually the wind over the fog is going a different direction and carries the ship away, but you can't tell it's happening. If you do get into fog, it's most important to figure out exactly where you are and then to figure out your heading to get you where you want to go.

That was our situation. L130 was by Borkum, and from there we knew how to get to our airfield.

The fog didn't only get thicker, but also taller. At first, the top of the fog was 200 meters [6], but now it was 350 meters [7]. Over the fog it was a totally clear, beautiful blue sky.

"That's what we were missing," suggested the watch officer. [8]

Yes, it wasn't pretty. We could probably spend hours looking for the flirtatious landing field. According to what we could observe of the current weather situation, the fog was going to be around the entire night and wouldn't diminish until the morning, when the lovely sun with its warm beams would make it all disappear. The simplest solution, which we would have never tried, was to just land in the fog before dark, but instead we could wait in the air till the next day. But who could guarantee that the fog would be gone tomorrow? And, the most important detail, our gasoline reserves were running empty.

The ship had a long scouting mission behind it that had consumed the largest portion of our gasoline. Once, I had to touch down at a nearer airfield than my home airfield due to low oil. After that the LoA [9], whom I had the privilege to meet at that airfield, talked to me, and I promised myself never to end up floating around as a free balloon, and landing, or more correctly said, to have to land, or be forced to land due to running out of gasoline. Excluding that, even though we were all very excited to fly on our airship, we also liked to land every now and then, especially when our stomachs began to growl.

"And tonight there will be potato pancakes, Mr. Over Lieutenant," said the watch officer very sadly, realizing he wasn't going to be able to get any.

"We'll get down." I tried to calm him. "No one has ever really been stuck in the air!"

The ship moved to 500 meters [10] altitude. According to our table settings [11], we must be about 20 kilometers [12] from the airfield. We have to pierce the fog if we are to get our orientation.

The ship was weighed off, and we made it a little "heavy", it wouldn't work any other way in the cold, denser air in the fog. [13]

"All engines half forward. Slowly descend to 100 meters! [14]"

The bow lowered and the ship slowly submerged itself into the "washroom". How this term came to be, the reader can probably imagine. The view in the fog and in the washroom are the same- you can't see anything.

It all depends on the forward gondola. Everyone looks down to see if they can spot even the smallest patch of earth.

"100 meters!",[14] reported the signal mate at the ruddervator.

"Lower to 75 meters!" [15]

Just thicker fog.

Now I have some of the engines shut off so the ship will slow down a bit. Very slowly we feel out for the ground below us.

At thirty meters [16] we could finally make out the ground. It was a forest. But which forest? In fog every forest looks the same.

"Two chimneys, straight ahead," the helmsman called out.

"Ruddervator hard up. One pant in front." [17]

In that same moment, a mass of water poured out of the ballast pant, the front of the ship got lighter and the nose shot up.

That was the brickyard west of the airfield. We were out four kilometers away. [18]

By dumping the forward pant of water, the ship was trimmed, naturally. People were sent from the back of the ship to the front, to trim the ship correctly.

Now we had to go down again.

The forward motor was completely shut off so that its noise didn't disturb us. The landing crew have probably heard it and are probably all in position with the alarm whistle going off.

Straight off to the starboard side, a fog bell is ringing.

The ship had just flown by the landing field. We've been lucky so far in that none of the hangars have gotten in our way. It's no fun when that happens. Those lousy hangars never get out of the way. A small chimney is infinitely easier to avoid than a giant hangar.

Alright, let's try another landing run.

We start circling about to the rear.

This ship is back on a landing approach. We need to spot the train tracks, they're right next to the landing field.

The ship is only 25 meters [19] high now. Below is a pasture with livestock. The spooked horses gallop and book it out of there [20] as soon as they see our giant bird overhead. The fat ones [21] trott first then work into a gallop. Only a couple stay standing and watch the sight above them in stoic calm. They seem to already know what the noise is.

There is a small farm below us. We're bound to get to the trainline soon. Chickens, ducks, and geese race excitedly to the barn and into their stalls. The small children come running out to the yard and scream at the top of their lungs, "hurra". There are the tracks, and behind them the fence and the airfield! We can already hear the ground crew shouting and whistling. "All engines stop!- descend!"

The ship sinks till about 10 meters [22] over the ground. We can see the people and they are running to us.

"Both engines, full power, reverse!" The ship slows. It slowly sinks the last ten meters [22], till the people have grabbed the mooring lines and the gondola.

"Shut off all engines!"
"
The officer of the ground crews greeted me with, "It was high time that you came down here. It was getting pretty dark and the fog was just getting thicker."

About fifteen minutes after the ship was safely in the hangar, it was so dark and the fog was so thick that landing could not have been considered.

Instead of sitting comfortably in the chow hall and enjoying potato pancakes, we would have had to stay in the sky all night.

And more importantly we didn't run out of gasoline!

* * *

About two hours had passed, when a runner came and reported to me: "The LoA [9] is summoning Mr. Over Lieutenant."

"Reporting as ordered."
"Did you see L19 while you were out? I haven't received any communications from them in the past two hours and I'm worried about that ship."

As I answered no, I received oral orders to prepare my ship, so that I could immediately take off and look for L19 at sea.

Nothing more came of that. The fog had gotten thicker, and the wind was blowing noticeably from the southeast. [23]

According to its last radio report, the ship was close to the coast of Holland. Then our connection with them suddenly ended.

Immediately, the Fleet commander sent torpedo boats and light [24] combat units out to sea to look for L19. But none of those vehicles reported back that they had seen them. The weather was also too unfavorable. The fog was already on the sea.

The next morning we got our first news. It came from Holland and said that an airship, traveling at low altitude in the fog around Holland had been shot down for violating neutrality.

That could only be L19.

The next day, an English report, vague and incomprehensible like all of Reuter's news reports, stated that the fishing boat "King Stephan" saw an airship floating on the North Sea and were unable to save the crew.

That was pretty much everything there was to know. Certainly, the ship had been lost. But we still hoped the crew could be saved.

"Why was the fishing boat unable to save them?"

About a month later, the shroud of secrecy was lifted.
The Swedish yacht "Stella Smoegen" fished out a message in a bottle off the western coast of Sweden. It was from L19.

Soon after another was found.
The following letters from the crew to their homes clarified the entire event all at once:

Bottle-Field postcard
Mrs. Mary [25] Uhle, Wilhelmshaven, #40 Boersen Street.

OverMate Uhle, Naval Airship L19, at Terschelling, Norwegian Street, February 1, 1916.

North Seas

February 1, 1916
My beloved wifey, my beloved laddie!

We were coming back from a big mission. We had bad luck with the engines. Now we're only flying at 100 meters. [26] Any minute now, we could be blown into the water. As God wills is the only thing we can think about when it comes to our rescue. Should God lead things differently, keep yourselves healthy. Receive my last greetings and kiss and keep your beloved Papa in your loving memories.
I have your picture with me.

Field-postcard

Mrs. Grete Baumann Stammbach,
                      Bavaria, Oberfranken, Germany

Departed Over-Machinist-Mate Georg Baumann,
Naval Airship L19, caught in peril at sea February 1, 1916, at 4 in the afternoon.

Dear Gretel and children!

I currently find myself in great danger. My ship and I are in the water. Beloved Gretel, I am hoping for rescue to the very last minute. If things are meant to be otherwise, then it is God's will. Faithful even in death and kiss you and the children with all your heart for your true Georg.

---

2. Message in a Bottle
Corvette Captain Strasser, Nordholz (Lehe).

With fifteen men on the platform and upper ridge in about 3 degrees [27] over the floating body of the L19, I attempt one last report. Three engines broken down, a light headwind on our way home delayed our arrival and brought us into fog, close to Holland, where I received lively fire. It was difficult, with three engines breaking down at the same time.

February 2, 1916, around 1 o'clock in the afternoon, is probably our last hour.

    Loewe.

Mrs Loewe, Luebeck, #13 Hansa Street
                            February 2, 1916, afternoon around 12 O'clock.
In my last hour on the platform in the presence of my people in three degrees [27] I have long been thinking of you. Thank you for everything. Nurture [28] our children!
                            Your Odo.

Mrs. Mary [25] Uhle,
            Wilhelmshaven, #40 Boersen Street

Departed Over Mate Uhle, Naval Airship L19,
            North Sea February 2, 1916, 12 O'clock midday.
            North Sea, Wednesday, February 2, 1916
My beloved wifey!

After we've fought with the sea for thirty hours, our last hour is come. We have said our prayers and so I turn you and Walter over to "God". Live well.
            Your Papa.

———

Pastor Braunhof, Rinteln by Hannover
            North Sea, February 2, 1916
Two days and two nights we've drifted. No help. Greetings to you. An English steamer didn't want to save us.
            Erwin

———

My beloved Ava and Mother!
It is 11AM February 2. We're all still alive, but have nothing to eat. This morning a fishing steamboat, an English one, was here. They didn't want to rescue us. It was called "King Stephen" from Grimsby. Our courage is depleted, the storm is strengthening. Yours, even in heaven, your thoughtful Hans.
About 11:30, we prayed together and said our goodbye's.
            Your Hans

———

These messages in bottles proved the fate of the glorious end of L19, its heroic crew, and the shame of the "King Stephan".[29]

The crew of L19 fell in difficult service to the Fatherland. The crew of the "King Stephan", whom we later captured, still live.
Losses in war are to be expected, even in our Naval Airship service. That can not stop our weapon from exactly following our highest warlords' orders, that we have been given, constantly working with all our might to further the assignments given:

For the Fatherland,
For Kaiser and Reich!

1- Below is a map of Friedrichshafen on Lake Constance. In German it is called Bodensee (Floor Lake), but for some reason in English we call it Lake Constance. It is on the border between Switzerland, Germany, and Austria. My borders are not exact. The exact borders are in grey and hard to see. I used higher contrast colors that are fairly close to the actual borders.

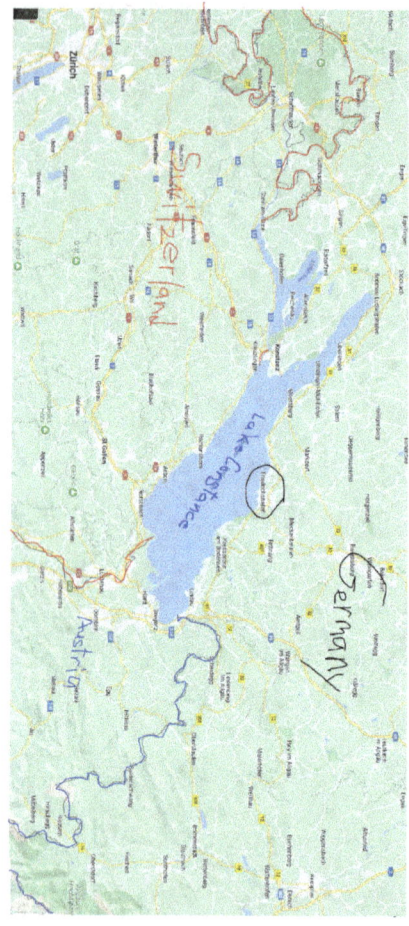

2- Old haunts as in places they used to hang out in.

3- I find it funny that the year when Count Zeppelin's ship got stuck on a pear tree, and the ship number, were both censored. I guess since it had all happened less than 15 years before this book was published, I

97

can believe they were being cautious with the development of cutting edge technology. Suffice to say, anyone can look it up now.

4- Darmstadt in Germany.

5-Rounders is a game related to cricket and baseball. I have never heard of it before, so I suspect it has fallen out of style in Germany.

6- About 656 feet

7- About 1148 feet

8- This is German facetious humor. As in, the situation wasn't bad enough, they needed fog to make it a bad situation.

9- Leader of Airships, the German abbreviation was FdL. Footnote 1 of Chapter 1 goes into detail about the title Leader of Airships.

10- About 1640 ft

11- I believe this is an old German idiom meaning something like "according to how we do things"

12- 12 miles

13- When an airship is in a cold environment, it can lift more than in a hot environment. Moving into colder air would make it harder to descend. They probably vented gas to get the weight of the ship right.

14- About 328 feet

15- 246 feet

16- 98 feet

17- The WW1 zeppelins had cloth water ballast tanks that looked like a giant tied off pant leg. So they called them pants.

18- 2.5 miles

19- About 82 feet

20- The German idiom was "tear away". I know "tear" is also used idiomatically in English for running really fast away from something, but this idiom felt more inline with what the author meant.

21- In German the author used the term "the turnips that fall hard" as slang for the "fat ones".

22- About 32.8 feet.

23- This would blow the fog out to sea, where they were trying to search.

24- Light as in not heavy.

25- In German, they add the suffix -chen (pronounced "shin") to names to make them more endearing. We do this in American English by adding y at the end, like Dan VS Danny, Bob Vs Bobby, Kate VS Katey,

99

or John vs Johny. The name here used the German suffix, but it can't be done in English because the name is Mary. It already ends in y.

26- About 328 feet.

27- About 37 degrees F

28- The German word used could have also been translated as "raise". I used nurture because it felt more fitting for the circumstances.

29- The story of L19's last mission was included in Rolf Marben's book, "Knights of the Air". He said he had access to the original messages found in the bottles and communications from L19, and that he wrote his best guess as to what happened. I was very happy to find that this book included what is supposed to be the original messages from the bottles. I know that this could very well be a propaganda effort and misinformation, but considering the WW1 Germans admitted to the Zimmerman telegram, I'm inclined to believe they couldn't lie even if it was for their own good. Unfortunately, if you want to verify any of this with original military documents, the vast majority of German WW1 airship operation records were confiscated by the British at the end of WW2 and have yet to be returned as of 2022. This is what the German Federal Archives told me when I tried to research their records regarding WW1 airship operations.

Advertisements:

My ride to Lille
from
Cavalry Master Prince Karl Wrede

How was Lille captured with just three men? We read the story a few months later in short newspaper stories. This book is the first account from the dashing commander of the troops that captured the city, Cavalry Master Prince Karl Wrede, of the glorious adventure that he and his two companions rode to. Shortly thereafter, the King of Bavaria awarded Prince Wrede the Military Order of Max Joseph, the highest military honor in Bavaria. Lille has been ours for a while now, but even in these decisive days, the Allies million men army have been futilely trying to find their way back. Earlier it only took three confident German soldiers. This book presents the past record of the daring the Prince and his companions encountered and the greater strength they had to use to deceive the commander of the city. After reading all about it in these pages, it will seem like a miracle, that these three German lancers [1] who captured, could confront a spies death and be surrounded by lurking danger, that they left Lille unharmed, and later officially and finally occupied the city with the large German army.

Price 1 Mark
Eckhart-Verlag A-G. Berlin SW 68
Lindenstrasse 105

1-The word used here was Parlancentaere. I couldn't find an exact translation, but my gut tells me it's lancers.

The Child and the War

Playground Rhymes[1], Essays, Drawings and Illustrations of Children from ages 5 to 11, collected and compiled by Max Schach

This collection shows an honest disposition with fresh humor how the children of our great events relate to them. Many of that which they coincidentally experience, is an expression of enemy sentiment, which they see differently. It's not that these depictions of slaughter, heroes and prisoners, from war, victory, and danger are correct. How deeply do the youngest children of our times think about these things, such as, "What is a soldier?" Or how excited they get when asked to say something about Hindenburg. Next to the precious impressions of the youngest children the nine and ten year olds seem experienced spectators. These pages reflect this in colorful childish inspirations, sensitivities, and thoughts. —Max Schach, who gathered and compiled these materials, in his introduction states this is a book, "written by the children for the adults". You read from the first to the last page with comfort, and enjoy the funny inspirations of Paul Simmel and Fritz Wolff, who in turn present whimsical illustrations, how the child thinks about the war. [2]

Price 2 Mark
Eckhart-Verlag A-G. Berlin SW 68
Lindenstrasse 105

1- Playground rhymes at the time period in Germany changed frequently and were a reflection of the children's understanding of current events. This was even depicted in the black and white German movie "M". The playground rhymes I grew up with in America were passed down and had no connection whatsoever with current events.

2-In German the last bit was able to be a play on words that included the title of the book. Literally it would have read: how they themselves "the child and the war" think. There was no way to keep the word play and the meaning, so I translated the meaning and tried to explain the word play.

The Circus Folk of our Enemies

With a forward by Leo Frobenius

128 pages of pictures (gravure) [1] with about 150 pages of text, elegantly bound

---

Of the many endorsements in the many newspapers, we reprint the 42nd one, from the "New Viennese Daily" from February 17th 1917: "An interesting picture book, with text written by Leo Frobenius .....It presents 128 gravure.....the unbelievable manifestations of the colored French and English shown here is the most perfectly complete ethnical collection......this book gives a good understanding of the revolting and unscrupulous efforts, that they strive towards our extinction. [2]. In any case, this document, carefully prepared with uncanny understanding presents a collection that is simply irrefutable eternal shame for those that claim to fight for civilization and freedom…

Price 2 Mark
Eckhart-Verlag A-G. Berlin SW 68
Lindenstrasse 105

1-A gravure is an image produced from etching a plate through an intaglio process and producing a print from it.
2- The German word used could also have been translated as destruction, but this felt more in line with the tone of the advertisement.

Over the Enemy
(An Airplane Book)[1]
From
Chief Otto Lehmann [2]

Among the many war books of our day, a special place is reserved for this courageous pilot. With this daring Officer's freshness, and compelling naturalness, the story of his daring flights are told. You will experience the excitement of the colorful multitudes of airplane pilot's existence in service to the Fatherland. The author experienced the hot fighting in the East that has been presented so clearly to us. From the Rhein, where the first test flight of the apparatus happened, the book flies into the action on the Eastern border of our Empire, and high over the enemy, the Chief can see the battlefield that is hidden from the other soldiers' view. You'll read with breathless excitement the beautifully presented snowstorm that this warrior entered, to meet his enemy high in the mountainous clouds. Every step of the dangerous work in the airplane service is presented with full visuals by Otto Lehmann, and brings to the reader the insignificant and significant experiences of a pilot, the days of hot fighting and the small hours of rest in their entirety. Clear, captivating pictures strengthen the impressions that come from this unique book.

Price 2 Mark
Eckhart-Verlag A-G. Berlin SW 68
Lindenstrasse 105

1-That they had to specifically say this was an airplane book, not an airship book, shows how prevalent airships were in the German WW1 public's consciousness.

2- There was also an airship pilot whose last name was Lehmann, who also had a bit of celebrity at the time. Probably another reason they clarified this was an airplane book.

Print: Albert Paul & Co. GmbH Berlin C. 25

Imperial German Navy Enlisted Rank Structure

	German Name	English Meaning	
1	Matrose	Seaman	*grunts*
2	Obermatrose	Over Seaman	
3	Maat	Mate	Petty Officers *experienced grunts*
4	OberMaat	Over Mate	
5	Vize-Wachtmeister	Vice Watch Master	
6	Wachtmeister	Watch Master	
7	Deckoffizier	Deck Officer	Warrant Officers *very experienced grunts*
8	Oberdeckoffizier	Over Deck Officer	
9	Deckoffiziere als Offizier-Stellvertreter	Deck Officer Serving as an Officer	
7	Vize-*	Vice-*	If the warrant officer had a trained specialty, they would be called by that specialty, such as vice-helmsman, over-engineer, etc…
8	*	*	
9	Ober-*	Over *	

106

Imperial German Navy Officer Rank Structure		
	German Name	English Meaning
1	Deckoffizier-Leutnant	Deck Officer, Lieutenant
2	Leutnant Zur See	Lieutenant at Sea
3	Oberleutnant zur See	Over Lieutenant at Sea
4	Kapitaenleutnant	Captain Lieutenant
5	Korvettenkapitaen	Corvette Captain
6	Fregattenkapitaen	Frigate Captain
7	Kapitaen zur See	Captain at Sea
8	Konteradmiral	Like an Admiral
9	Vizeadmiral	Vice Admiral
10	Admiral	Admiral
11	Grossadmiral	Great Admiral

www.ingramcontent.com/pod-product-compliance
Lightning Source LLC
Chambersburg PA
CBHW070320220526
45465CB00013B/1507